Math Challenge the Singapore Way

Grade 5

Copyright © 2012 Marshall Cavendish Corporation

Published by Marshall Cavendish Education

Marshall Cavendish Corp.
99 White Plains Road
Tarrytown, NY 10591
Website: www.marshallcavendish.us/edu

Originally published as Skills in Problem-Solving Maths Copyright © 2002 Times Media Private Limited, Copyright © 2003, 2010 Marshall Cavendish International (Singapore) Limited

All rights reserved.

No part of this publication may be reproduced, stored in a retrieval system or transmitted, in any form or by any means, electronic, mechanical, photocopying, recording, or otherwise, without the prior permission of the copyright owner. Request for permission should be addressed to the Publisher, Marshall Cavendish Corporation, 99 White Plains Road, Tarrytown, NY 10591. Tel: (914) 332-8888, fax: (914) 332-1888.

Other Marshall Cavendish Offices:
Marshall Cavendish International (Asia) Private Limited, 1 New Industrial Road, Singapore 536196 • Marshall Cavendish International (Thailand) Co Ltd. 253 Asoke, 12th Flr, Sukhumvit 21 Road, Klongtoey Nua, Wattana, Bangkok 10110, Thailand • Marshall Cavendish (Malaysia) Sdn Bhd, Times Subang, Lot 46, Subang Hi-Tech Industrial Park, Batu Tiga, 40000 Shah Alam, Selangor Darul Ehsan, Malaysia.

Marshall Cavendish is a trademark of Times Publishing Limited

ISBN 978-0-7614-8031-0

Printed in the United States
135642

Introduction

Math the Singapore Way is a term coined to refer to the textbook series used in Singapore schools. Math the Singapore Way focuses on problem solving, given that is it based on a curriculum framework that has mathematical problem solving as its focus. Math the Singapore Way also focuses on thinking, given that the Singapore education system is driven by the *Thinking Schools, Learning Nation* philosophy. Math the Singapore Way is also based on learning theories that provide clear directions on how mathematics is learned and should be taught.

Singapore mathematics textbooks, initial teacher preparation, and the subsequent professional development for teachers are based on helping teachers understand what to teach in mathematics and how to teach it.

While these books are not part of the formal classroom program, they provide selected groups of students with a necessary challenge. Some students complete basic materials easily and enjoy moving on to more challenging tasks. These books are written with that goal in mind. Such learning materials, if they are prepared consistently with the fundamentals of Math the Singapore Way, place an emphasis on problem solving and provide students with various strategies to solve problems, including the use of visuals. It should be noted that good practice is not a matter merely of random repetition. Learners must be challenged through sets of carefully crafted problems. Good practice consists of careful variations in the tasks learners are given.

I hope this series of books is able to provide the necessary help for learners who need to be challenged beyond basic concepts and skills.

Yeap Ban Har
Marshall Cavendish Institute

Preface

Math Challenge the Singapore Way was written to provide students with the practice needed to excel in Math. The problems have been designed in accordance with the latest Math syllabus used in Singapore.

Problems have been differentiated into two levels in each exercise. These exercises test students on their understanding of the mathematical concepts. Questions that allow calculator use are indicated with a 🖩. The more difficult questions are highlighted with a Math Buddy icon 🐕.

Worked examples are also provided at the start of each exercise to help students understand the mathematical processes involved.

With this book, students will be exposed to a wide variety of problems that will help them understand math and take their school and state exams with greater confidence.

Notes pages are provided at the back of the book for those who require more space to work out the solutions to the problems.

Contents

Exercise 1	Whole Numbers	1
Exercise 2	Fractions	15
Exercise 3	The Area of Triangles	26
Exercise 4	Ratio	33
Exercise 5	Geometry	43
Exercise 6	Decimals	50
Exercise 7	Percentage	65
Exercise 8	Average	73
Exercise 9	Volume	78
Test Yourself 1		87
Test Yourself 2		92
Answers		97

Exercise 1

Whole Numbers

Notes:

ROUNDING OFF TO THE NEAREST THOUSAND & ESTIMATION

To round off a number to the nearest thousand, we look at the hundreds place. If the number is 5, 6, 7, 8, or 9, we round it up. If it is 1, 2, 3, or 4, we round it down.

7,521 is 8,000 when rounded off to the nearest thousand.
7,388 is 7,000 when rounded off to the nearest thousand.

To check whether answers are likely, we estimate by rounding off 4-digit numbers to the nearest thousand.

$2,422 + 3,802 \approx 2,000 + 4,000$
$\qquad\qquad\quad = 6,000$

$9,023 - 6,241 \approx 9,000 - 6,000$
$\qquad\qquad\quad = 3,000$

$3,156 \times 9 \approx 3,000 \times 9$
$\qquad\quad = 27,000$

To estimate $6,158 \div 8$, we choose a number that can be divided by 8 exactly.

$6,158 \div 8 \approx 6,400 \div 8$
$\qquad\quad = 800$

ORDER OF OPERATIONS

Work from left to right when adding and subtracting.

$99 - 23 + 21 = 76 + 21$
$\qquad\qquad\quad = 97$

Work from left to right when multiplying and dividing.

20 × 3 ÷ 4 = 60 ÷ 4
= 15

Working from left to right, carry out multiplication and division before addition and subtraction.

43 + 21 ÷ 7 = 43 + 3
= 46

When there are brackets, carry out the operations in the brackets first.

(450 + 250) ÷ 10 = 700 ÷ 10
= 70

Example 1

Jane and her two brothers had 240 stamps altogether. Jane had twice as many stamps as the sum of her two brothers' stamps. If each brother had the same number of stamps, how many stamps did each brother have?

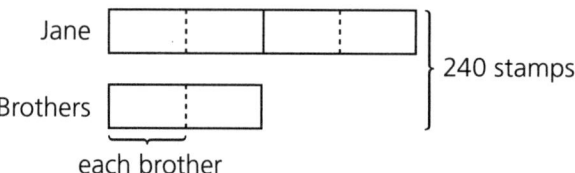

6 units → 240 stamps
1 unit → 240 ÷ 6
= 40 stamps
Each brother had 40 stamps.

Example 2

There are 2,470 children at a book fair. There are 948 more boys than girls. How many boys are there at the book fair?

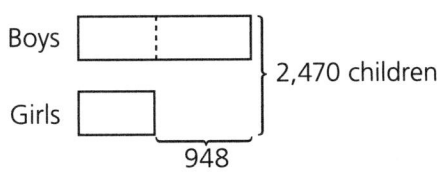

When there are brackets, carry out the operations in the brackets first.

$(2{,}470 - 948) \div 2 + 948 = 1{,}709$

There are 1,709 boys at the book fair.

Example 3

A jug and 6 mugs cost $9. If the jug costs $2 more than a mug, find the cost of the jug.

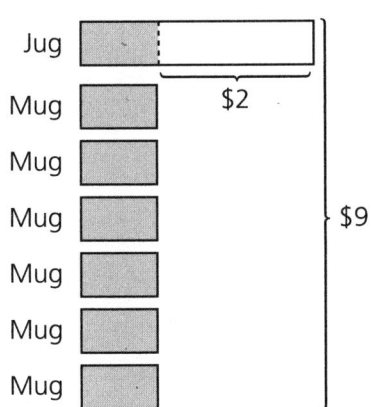

Cost of 7 mugs = $9 − $2
             = $7

Cost of a mug = $7 ÷ 7
             = $1

Cost of a jug = $1 + $2
             = $3

The jug costs $3.

Example 4

Minah had $200. She gave her brother $20 and found that she had three times as much money as her brother did. How much money did her brother have at first?

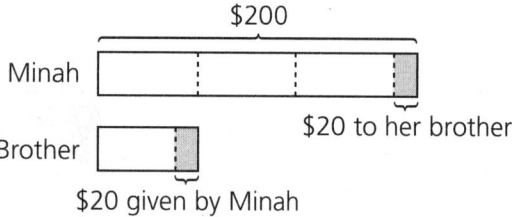

Amount of money Minah had left = $200 − $20
= $180

3 units → $180
1 unit → $180 ÷ 3
= $60

$60 − $20 = $40
Her brother had $40 at first.

Example 5

Mr. Li gave his two children some money. His son received twice as much money as his daughter. If his son spent $68 and his daughter received $66 more from his wife, they would have the same amount of money. How much did Mr. Li give his son?

1 unit → $68 + $66
= $134
2 units → 2 × $134
= $268
Mr Li gave his son $268.

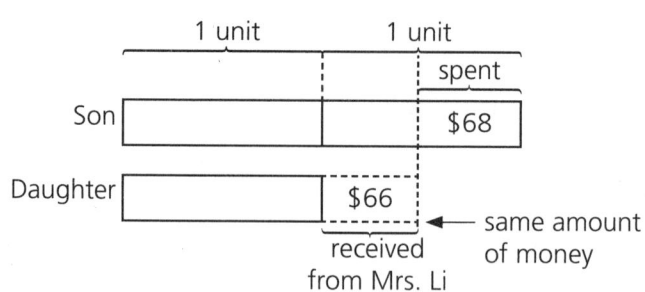

Example 6

A machine can produce 720 toys in 9 minutes.

(a) How many toys can it produce in 4 minutes?

$$9 \text{ min.} \rightarrow 720 \text{ toys}$$
$$1 \text{ min.} \rightarrow 720 \div 9$$
$$= 80 \text{ toys}$$
$$4 \text{ min.} \rightarrow 4 \times 80$$
$$= 320 \text{ toys}$$

It can produce 320 toys in 4 minutes.

(b) How long will it take to produce 900 toys?

$$720 \text{ toys} \rightarrow 9 \text{ min.}$$
$$1 \text{ toy} \rightarrow \left(\frac{9}{720}\right) \text{min.}$$
$$900 \text{ toys} \rightarrow 900 \times \frac{9}{720}$$
$$= 11\frac{1}{4} \text{ min.}$$

It will take $11\frac{1}{4}$ minutes to produce 900 toys.

Level 1 Work out the following problems.

1. Ms. Gomez bought 12 dozen pencils. She repacked them into packets of 10 each. How many packets were there? How many pencils were left over?

2. In a school, there were 53 more boys than girls. If there were 642 boys, what was the enrollment of the school?

3. A toy bear costs $39. A toy dog costs $6 more. Find the total cost of 2 toy bears and 3 toy dogs.

4. A painting costs $498. It costs twice as much as a vase. Mr. Wilson bought both the painting and the vase. He gave the cashier $1,000. How much change did he get back?

5. A study table and a chair cost $168. The table costs $110 more than the chair. Find the cost of the chair.

6. A dress costs twice as much as a skirt. Ms. Blackwell bought 2 dresses and 2 skirts. She paid a total of $150. Find the cost of each skirt.

7. An executive assistant can type 280 words in 4 minutes. How many words can he type in one minute?

8. Mrs. Barreda can make 25 tamales in one minute. How many tamales can she make in 12 minutes?

9. Ms. Greenwood bought 5 apples for $2. How many apples can she buy with $10?

Level 2 Work out the following problems.

1. A store owner bought 350 rulers at 40¢ per ruler. If the cost of each ruler had been 5¢ less, how many more rulers could he buy with the same amount of money?

2. Mrs. Harjo bought 3 pears and 4 oranges. Mrs. Tsotigh bought 3 oranges and 4 pears. Mrs. Tsotigh paid 30¢ more than Mrs. Harjo. If Mrs. Harjo paid $4.40, how much did 2 pears cost? Give your answers in cents.

3. Container A had 15 liters of water. Container B had 5 liters of water. When an equal amount of water was poured into both containers, Container A had twice the volume of water as Container B. What was the least amount of water that was poured into each container?

4. Two brothers received the same amount of money from their mother. After Tom had spent $12, and Terry had spent $4, Terry was left with twice the amount of money that Tom had. How much money did each boy receive at first?

5. 35 ounces of flour and 25 ounces of sugar cost $6. Lena bought 210 ounces of flour and 275 ounces of sugar for $56. How much did 25 ounces of sugar cost?

6. Brian had $5 less than Grace. If Brian gave Grace $4, she would have twice as much money as Brian. How much money did each of them have at first?

7. Tap A can fill a 30-ounce container in 1 minute. Tap B can fill only half of the same container in 1 minute. How long will it take for both taps to fill the container at the same time? Give your answer in seconds.

8. Mail to the United Kingdom is sent at the following postal rates:

> Up to 20 g — 75¢
> Next 30 g — 20¢ per 10 g
> Next 50 g — 40¢ per 10 g
> Next 100 g — 50¢ per 10 g

Ms. Baker wants to mail a 60-gram letter and a 100-gram letter to her 2 friends in the United Kingdom.

(a) How much will it cost her to mail both letters as 2 separate pieces of mail?
(b) How much more will it cost her to mail both letters together as 1 piece of mail?

9. Ali and Jorge have the same number of crackers. If Ali eats 12 crackers every day and Jorge eats 3, Jorge will have 45 left when Ali has finished all his crackers. How many crackers does Jorge have at first?

10. The cost of 2 identical computers and 5 identical TVs is $28,322. A TV costs three times as much as a computer. Find the total cost of a TV and a computer.

Exercise 2

Fractions

Example 1

Sally had $\frac{2}{3}$ pounds of flour. She used $\frac{1}{8}$ pounds of flour to bake muffins. She then used the remaining flour to bake 13 cookies. How much flour was used for each cookie?

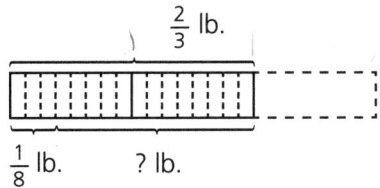

$\frac{2}{3} - \frac{1}{8} = \frac{13}{24}$ lb.

Sally had $\frac{13}{24}$ pounds of flour left.

$\frac{13}{24} \div 13 = \frac{1}{24}$ lb.

$\frac{1}{24}$ pounds of flour was used for each cookie.

Example 2

Purvi has some ribbon. She cuts off 8 pieces of ribbon, each $2\frac{4}{5}$ meters. She has $27\frac{3}{5}$ meters of ribbon left. How much ribbon did Purvi have at first?

$2\frac{4}{5} \times 8 = 22\frac{2}{5}$ m

Purvi cut off $22\frac{2}{5}$ meters of ribbon altogether.

$22\frac{2}{5} + 27\frac{3}{5} = 50$ m

Purvi had 50 meters of ribbon at first.

© 2012 Marshall Cavendish Corporation

Example 3

Stamps from the United States made up $\frac{3}{10}$ of Steven's stamp collection. The remaining stamps were foreign. He gave $\frac{2}{7}$ of his foreign stamps away. What fraction of his stamp collection did he give away?

$1 - \frac{3}{10} = \frac{10}{10} - \frac{3}{10} = \frac{7}{10}$

$\frac{7}{10}$ of Steven's stamp collection were foreign stamps.

$\frac{2}{7} \times \frac{7}{10} = \frac{1}{5}$

He gave away $\frac{1}{5}$ of his stamp collection.

Example 4

Clarissa took $3\frac{1}{5}$ hours to finish reading a book. Brian took $\frac{2}{3}$ hours less to finish reading the same book. How much time did it take them to finish reading both books altogether?

$3\frac{1}{5} - \frac{2}{3} = 2\frac{8}{15}$ hours

Brian took $2\frac{8}{15}$ hours to finish reading the book.

$2\frac{8}{15} + 3\frac{1}{5} = 5\frac{11}{15}$ hours

They took $5\frac{11}{15}$ hours to finish reading their books altogether.

Example 5

Bethany had $30. She spent $\frac{1}{6}$ of it on Monday and $\frac{2}{5}$ of it on Tuesday. How much money did she have left?

$$\frac{1}{6} + \frac{2}{5} = \frac{5}{30} + \frac{12}{30} = \frac{17}{30}$$

She spent $\frac{17}{30}$ of the money altogether.

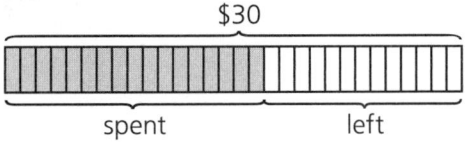

30 units → $30
1 unit → $30 ÷ 30
= $1
13 units → 13 × $1
= $13
She had $13 left.

If I had spent 17 units of the money, I would have (30 − 17) = 13 units left.

Example 6

Gloria's mother gave her some money. Gloria spent $\frac{1}{4}$ of it on food and $6 on a toy. She had $\frac{3}{8}$ of her money left. How much money did Gloria's mother give her?

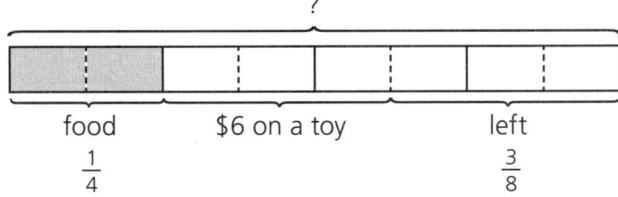

3 units → $6
1 unit → $6 ÷ 3
= $2
8 units → 8 × $2
= $16
Gloria's mother gave her $16.

Example 7

In an orchard, there were 18 rows of trees. In each row, there were 15 trees. $\frac{1}{9}$ of the trees were pecan trees. Of the remainder, $\frac{2}{5}$ were orange trees. The rest were lime trees. How many lime trees were there?

18 × 15 = 270 trees
There were a total of 270 trees in all.

First, find the total number of trees.

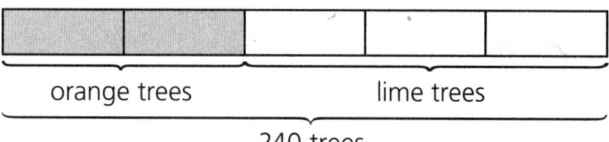

9 units → 270 trees
1 unit → 270 ÷ 9
       = 30 trees
8 units → 8 × 30
       = 240 trees
There were 240 orange and lime trees.

5 units → 240 trees
1 unit → 240 ÷ 5
       = 48 trees
3 units → 3 × 48
       = 144 trees
There were 144 lime trees.

Level 1 Work out the following problems.

1. Mr. Johnson had $14\frac{2}{3}$ pounds of flour. His wife had $10\frac{1}{4}$ pounds of flour. Together, they used $20\frac{5}{6}$ pounds of the flour to make some bread. How much flour did they have left?

2. A tailor bought some cloth. He used it to make 8 pairs of shorts. Each pair of shorts required $\frac{3}{4}$ yards of cloth. After that, he had $1\frac{1}{2}$ yards of cloth left. How many yards of cloth did he buy?

3. Mrs. Browning had 14 yards of cloth. She used $\frac{3}{5}$ of it to make a dress. How many yards of cloth did she have left?

4. Mrs. Behar spent $\frac{2}{5}$ of her money. If she had $15 left, how much money did she have at first?

5. A store owner had $6\frac{1}{3}$ pounds of sugar. His brother had $\frac{5}{6}$ pounds more than he had. What was the total weight of sugar they had altogether?

6. Boys make up $\frac{4}{7}$ of a school's enrollment. The difference between the number of boys and girls is 250. What is the enrollment of the school?

Level 2 Work out the following problems.

1. $\frac{3}{4}$ of Samuel's weekly allowance is $18. How much allowance does he get after 4 weeks if he gets the same amount every week?

2. Keith had $840. He spent $\frac{2}{3}$ of it on a camera and $\frac{5}{7}$ of the remainder on a tennis racket. How much money did he have left?

3. Lillian read a novel that was 420 pages long. She read $\frac{2}{7}$ of it on Saturday and $\frac{3}{5}$ of the remainder on Sunday. How many pages did she have left to read after that?

4. An orchard owner had 400 apples and 350 oranges. He threw away $\frac{1}{5}$ of the fruit because it was rotten. He sold 140 oranges and 180 apples. How many apples and oranges did he have left?

5. Helen had 23 postcards. Alice, who had 120 postcards, gave $\frac{5}{12}$ of her cards to Helen. Then, Jean gave Helen $\frac{2}{9}$ of her postcards. Now, Helen has 97 postcards altogether. How many postcards did Jean have at first?

6. Jack had some wooden boxes. He painted $\frac{1}{6}$ of them black, $\frac{5}{12}$ of them red, and $\frac{7}{24}$ of them blue. He painted the remaining boxes green.

   (a) How many more red boxes than green boxes were there? Write your answer as a fraction.
   (b) If 112 boxes were painted blue, how many boxes did he paint altogether?

7. Raymond bought an airline ticket to France at $\frac{4}{5}$ of the original price. If he spent $\frac{5}{9}$ of his money to buy the airline ticket, he would still have $1,680 left. What was the original price of the airline ticket?

8. A garment worker received an order to sew some garments. She sewed the same number of garments each day. After 4 days, she had sewn $\frac{1}{3}$ of the total number of garments needed. After another 3 days, she had 945 garments left to sew.

   (a) How many garments did she have to sew altogether?
   (b) What fraction of the total number of garments would she have sewn after 10 days?

Exercise 3

The Area of Triangles

Notes:

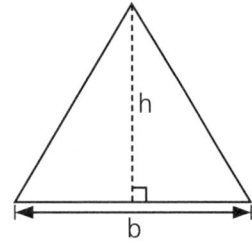 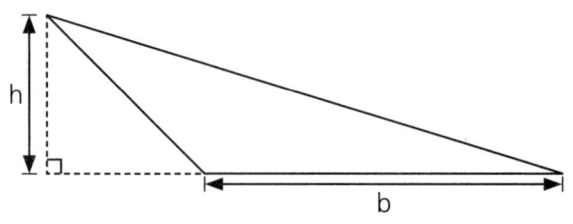

Area of triangle (A) = $\dfrac{\text{Base (B)} \times \text{Height (H)}}{2}$

(Units in cm², m²)

Height of triangle (H) = $\dfrac{2A}{B}$

(Units in cm, m)

Base of triangle (B) = $\dfrac{2A}{H}$

(Units in cm, m)

Example 1

The base of a triangle is 6 centimeters. Its height is 8 centimeters. Find its area.

Area of triangle = $\dfrac{B \times H}{2}$

$= \dfrac{6 \times 8}{2}$

$= \dfrac{48}{2}$

$= 24 \text{ cm}^2$

The area of the triangle is 24 square centimeters.

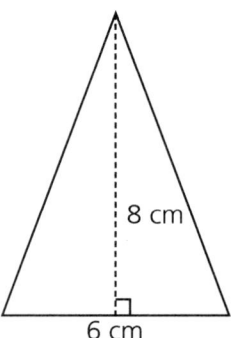

Example 2

In the figure below, AB is half the length of BE. Find the area of the figure.

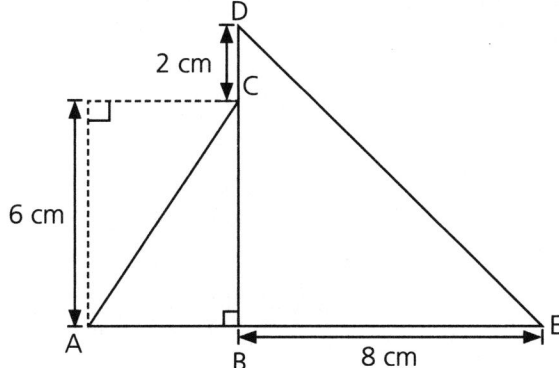

Length of AB = 8 ÷ 2
= 4 cm

Area of △ABC = $\frac{6 \times 4}{2}$

= $\frac{24}{2}$

= 12 cm²

Length of BD = 6 + 2
= 8 cm

Area of △BDE = $\frac{8 \times 8}{2}$

= $\frac{64}{2}$

= 32 cm²

Area of △ABC and △BDE = 12 + 32
= 44 cm²

The area of the figure is 44 square centimeters.

Level 1 Work out the following problems.

1. Mr. Aguillar made the signboard below using cardboard. Find the area of cardboard used to make this signboard.

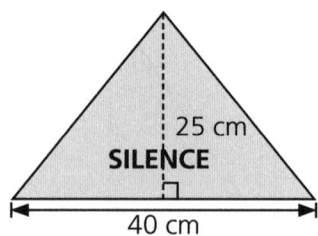

2. What is the area of the triangle below?

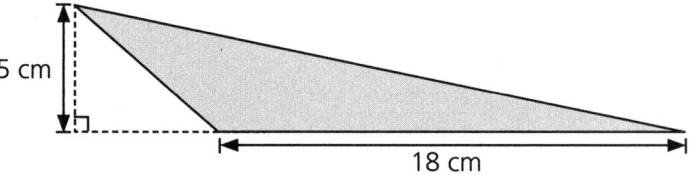

3. The area of triangle A is three times that of the triangle shown in the figure below. Find the area of triangle A.

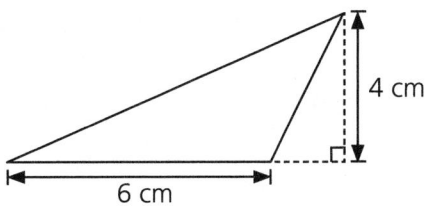

4. Joyce used a triangular piece of paper with a base of 24 centimeters and a height of 8 centimeters to cut out 2 smaller triangular shapes, each with an area of 19 square centimeters. Find the area of the remaining piece of paper.

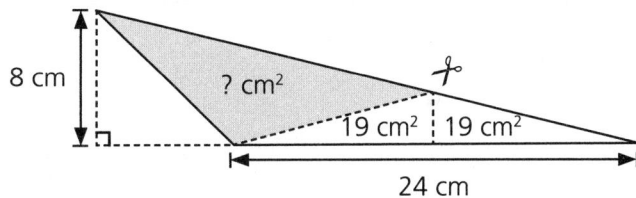

Level 2 Work out the following problems.

1. In the figure shown below, the length of BD is three times as long as the length of AC. Find the difference in area between triangle CBD and triangle ABC.

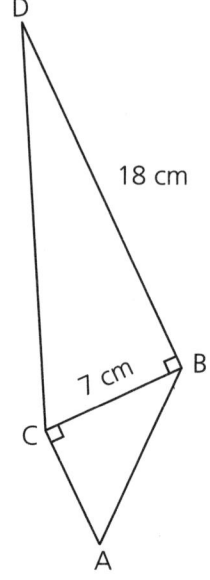

2. Triangle ABC was folded along AD as shown below. What is the area that is not covered by the folded area ADE?

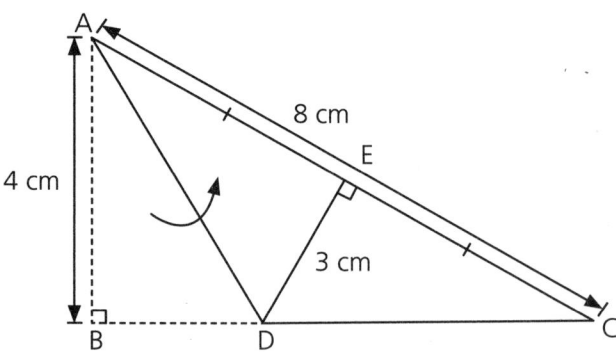

3. A and B are identical triangles. C and D are another pair of identical triangles. Find the total area of the 4 triangles.

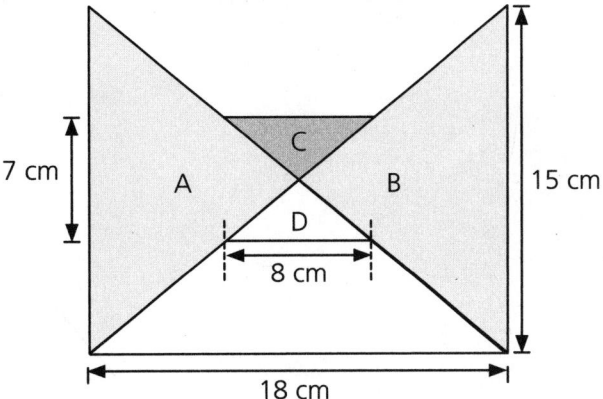

 4. The area of triangle ABE is $\frac{7}{2}$ of the area of triangle ACD. CD = DE and AC = 6 centimters. Find the area of triangle ABE.

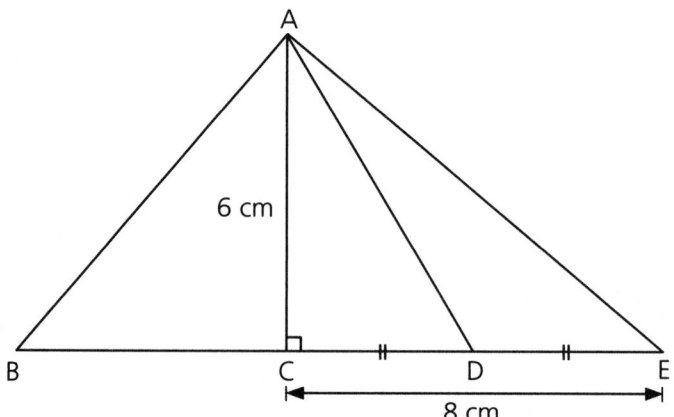

Exercise 4

Ratio

Example 1

Jenny and Sonja shared a sum of money in the ratio 3 : 4. If Sonja received $6 more than Jenny, find the total sum of money shared.

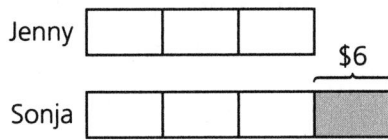

1 unit → $6
7 units → 7 × $6
= $42

The total sum of money shared was $42.

Example 2

672 pounds of dinner rolls are packed into two boxes in the ratio 5 : 3. The smaller portion is delivered to Store A and Store B in the ratio 3 : 4. How many more pounds of dinner rolls does Store B get than Store A?

5 + 3 = 8 units
8 units → 672 lb.
1 unit → 672 ÷ 8
= 84 lb.
3 units → 3 × 84
= 252 lb.

The smaller portion is 252 pounds.

3 + 4 = 7 units
7 units → 252 lb.
1 unit → 252 ÷ 7
= 36 lb.

Store B gets 36 more pounds of dinner rolls than Store A.

Example 3

Lionel and Paul shared some stamps in the ratio 2 : 5. If Paul had given 30 stamps to Lionel, they would each have the same number of stamps. How many stamps did Lionel have at first?

Before

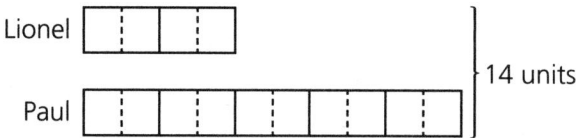

If they both had the same number of stamps, each would have 7 units of stamps.

After

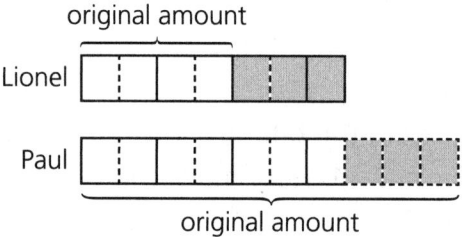

Paul would have given 3 units of stamps to Lionel.

3 units → 30 stamps
1 unit → 30 ÷ 3
        = 10 stamps
4 units → 4 × 10
        = 40 stamps
Lionel had 40 stamps at first.

Level 1 Work out the following problems.

1. Rafael has 24 marbles while Jimmy has 28 marbles. Show the number of marbles Jimmy has as a ratio of the total number of marbles they have altogether.

2. The cost of a calculator and that of a computer is in the ratio 1 : 9. If the computer costs $360, what is the cost of the calculator?

3. The ratio of the amount of salt in Sack A to the amount of salt in Sack B is 3 : 2. If Sack A contains 6 pounds more salt than Sack B, what is the total amount of salt in the two sacks?

4. In a class, the ratio of the number of girls to the total number of students is 2 : 5. If there are 14 girls in the class, how many boys are there?

5. Ivan and his brother shared some comic books in the ratio 1 : 4. His brother gave some of his books to Ivan. After that, they both had 10 books each. How many comic books did Ivan have at first?

6. Jane and her sister have $100 altogether. If Jane has $20 more than her sister, show the amount of money her sister has as a ratio of the amount of money Jane has.

Level 2 Work out the following problems.

1. Half of Sara's age is equal to a third of Paige's age. Find the ratio of Paige's age to Sara's age.

2. Two sisters, Liliana and Isabel, share a sum of money. If Liliana gives half of her share to Isabel, Isabel will have three times as much money as Liliana. Find the ratio of Isabel's share to Liliana's share at first.

3. In April, the salaries of Mr. Lillard and Mr. Sanders are in the ratio 5 : 8. If Mr. Lillard earns $2,310 more, his salary will be twice as much as Mr. Sanders'. Find both their salaries in April.

4. The ratio of the number of boys to the number of girls in a club last year was 3 : 4. This year, 22 girls joined the club and the ratio became 3 : 5. How many members were in the club last year?

5. Mr. Allen gave money to his wife and four sons in the ratio 3 : 5. His four sons had an equal share each. If his wife received $135,696, how much did each son receive?

6. The ratio of the number of oranges left unsold in Stall A at the farmers' market to those left unsold in Stall B was 3 : 7. Stall B had 24 more oranges left than Stall A. If Stall B sold another 6 oranges, what would be the new ratio of the number of oranges left in Stall A to those in Stall B?

7. The ratio of Ali's weight to Ben's weight is 6 : 7. The ratio of Charlie's weight to that of Ali's weight is 5 : 6. If Charlie's weight is 50 pounds, what is Ben's weight?

8. Pravin had some red and blue marbles. The ratio of the number of red marbles to the number of blue marbles was 2 : 5. He gave away 462 blue marbles in exchange for 172 red marbles. The difference between the number of blue marbles and red marbles is now 65. How many red marbles did he have at first?

Exercise 5

Geometry

Notes:

Angles

Angles on a straight line		The sum of all the angles on a straight line is 180°. $\angle x + \angle y + \angle z = 180°$
Vertically opposite angles		$\angle a$ is vertically opposite to $\angle b$. $\angle a = \angle b$
Angles at a point		The sum of all the angles at a point is 360°. $\angle x + \angle y + \angle z = 360°$

Properties of:

Parallelogram		(i) Opposite sides are parallel and equal. (ii) Opposite angles are equal. $\angle a = \angle c$ $\angle b = \angle d$
Rhombus		(i) Opposite sides are parallel and all sides are equal. (ii) Opposite angles are equal. $\angle a = \angle c$ $\angle b = \angle d$

Properties of:

Trapezoid	(trapezoid figure with angles a, b, c, d)	(i) One pair of opposite sides is parallel. (ii) Each pair of angles between the parallel sides adds up to 180°. $\angle a + \angle d = 180°$ $\angle b + \angle c = 180°$
Equilateral triangle	(equilateral triangle with angles a, b, c)	(i) All sides are equal. (ii) All angles are equal. $\angle a = \angle b = \angle c = 60°$
Isosceles triangle	(isosceles triangle with angles a, b)	(i) Two equal sides. (ii) Two equal base angles. $\angle a = \angle b$
Triangle	(triangle with angles a, b, c)	(i) The sum of angles in a triangle is 180°. $\angle a + \angle b + \angle c = 180°$

Level 1 Work out the following problems.

The following figures are not shown to scale.

1. Find ∠a given that ∠b = 62° and ∠c = 77°.

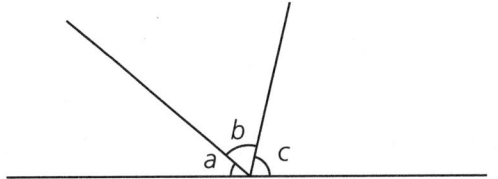

2. The diagram below shows angles at a point. Find ∠b.

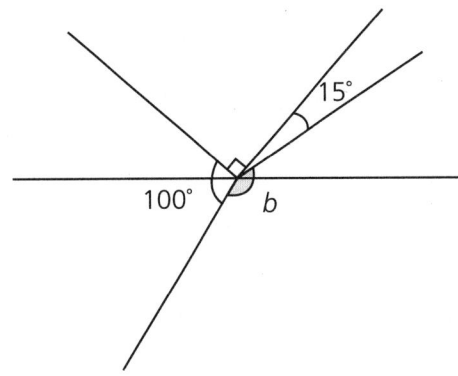

3. In the figure below, ABC, DBE, and GBF are straight lines. Find ∠c.

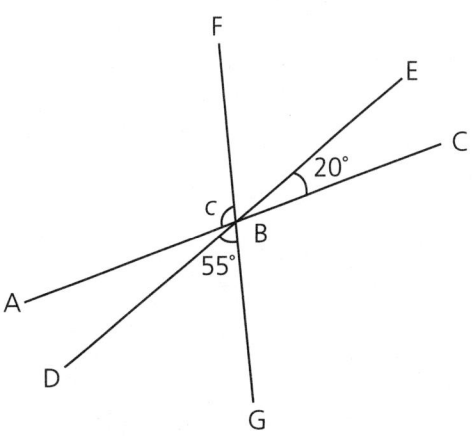

4. ABCD is a rhombus.

   (a) If AB = 10 centimeters, find the perimeter of the figure.
   (b) If DAB = 45°, find ∠DCB.

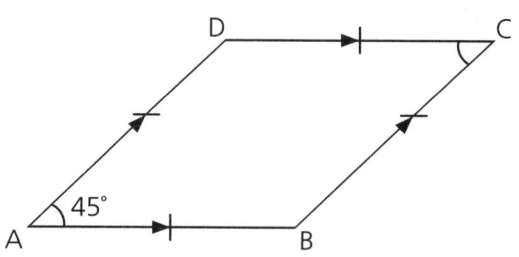

5. In PQRS, PQ // SR and PS // QR. PQ = SR and PS = QR.

   (a) Find the value of the sum of ∠SPQ and ∠PSR.
   (b) If ∠QRS = 65°, find ∠SPQ.

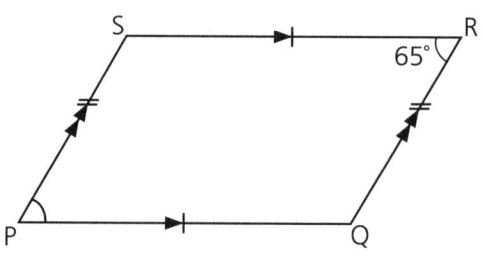

6. WXYZ is a trapezoid in which WX // ZY. Given that ∠YZW is 115°, find ∠ZWX.

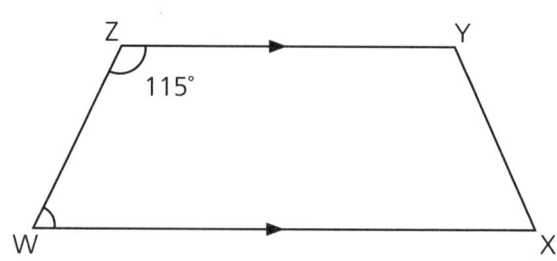

7. Using a ruler and a protractor, complete the figure below to form a rhombus.

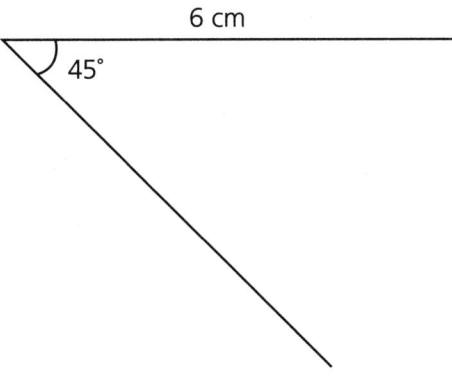

8. Using a ruler and a protractor, complete the figure below to form a parallelogram.

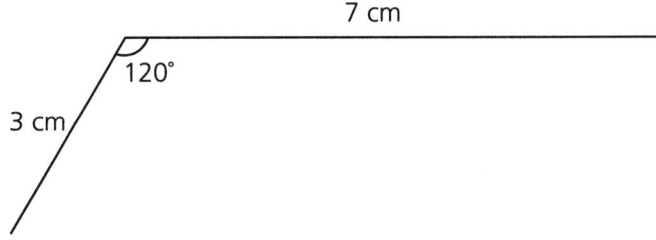

9. Using a ruler and a protractor, draw a triangle ABC such that AB = 4 centimeters, BC = 5 centimeters, and ∠ABC = 90°.

10. In the figure below, find ∠y.

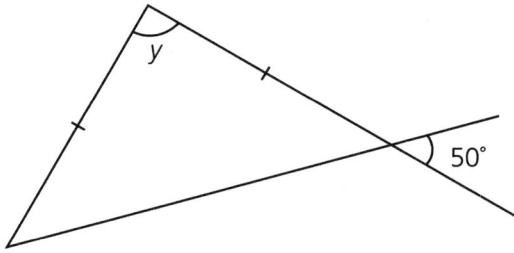

11. In the figure shown, ACD is an equilateral triangle and ABC is an isosceles triangle. If ∠DAB = 132°, find ∠ABC.

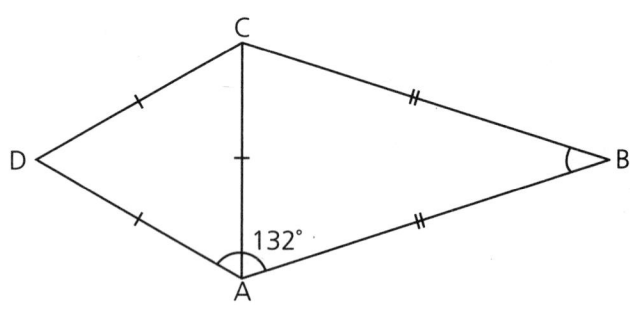

12. PQR is an isosceles triangle and PQ // TS. Find ∠c and ∠d.

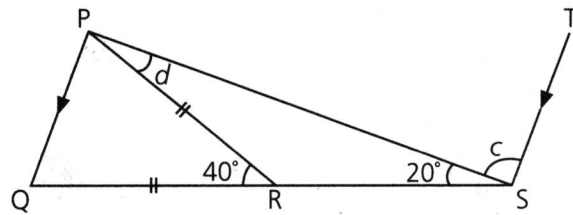

13. In the figure below, BT is a straight line and ABO is an equilateral triangle. Find ∠q and ∠s.

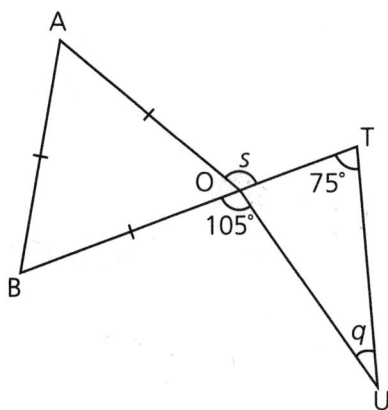

Exercise 6

Decimals

> **Notes:**
>
> ## MULTIPLICATION
>
> To multiply a decimal by 10, shift the decimal point 1 place to the right.
>
> 1.431 × 10 = 14.31
>
> To multiply a decimal by 100, shift the decimal point 2 places to the right.
>
> 1.431 × 100 = 143.1
>
> To multiply a decimal by 1,000, shift the decimal point 3 places to the right.
>
> 1.431 × 1,000 = 1,431
>
> ## DIVISION
>
> To divide a decimal by 10, shift the decimal point 1 place to the left.
>
> 735.4 ÷ 10 = 73.54
>
> To divide a decimal by 100, shift the decimal point 2 places to the left.
>
> 735.4 ÷ 100 = 7.354
>
> To divide a decimal by 1,000, shift the decimal point 3 places to the left.
>
> 735.4 ÷ 1,000 = 0.7354

ROUNDING OFF & ESTIMATION

To round off a number to 2 decimal places, we look at the thousandths place. If the number is 5, 6, 7, 8, or 9, we round it up. If it is 1, 2, 3, or 4, we round it down.

76.155 is 76.16 when rounded off to 2 decimal places.

To round off a number to 1 decimal place, we look at the hundredths place. If the number is 5, 6, 7, 8, or 9, we round it up. If it is 1, 2, 3, or 4, we round it down.

76.15 is 76.2 when rounded off to 1 decimal place.

To round off a number to the nearest whole number, we look at the tenths place. If the number is 5, 6, 7, 8, or 9, we round it up. If it is 1, 2, 3, or 4, we round it down.

76.1 is 76 when rounded to the nearest whole number.

To check whether answers are likely, we estimate by rounding off the decimal to the nearest whole number first.

$1.42 \times 14 \approx 1 \times 14$
$= 14$

Example 1

The weight of a table is 29.437 pounds. Estimate the weight of 5 such tables.

$29.437 \times 5 \approx 29 \times 5$
$= 145$ lb.

The weight of 5 such tables is about 145 pounds.

Example 2

The height of a wooden block is 12.28 centimeters. Estimate the height of 14 similar blocks stacked on top of each other.

$$12.28 \times 14 \approx 12 \times 14$$
$$= 168 \text{ cm}$$

The height of 14 similar blocks stacked on top of each other is about 168 centimeters.

Example 3

Mr. Clark had 15 sacks of rice, each with a weight of 6.25 pounds. He repacked them into sacks, each with a weight of 4 pounds. In the process of repacking, he lost 1.75 pounds of rice. How many 4-pound sacks of rice did he have?

$6.25 \times 15 = 93.75$ lb.
He had 93.75 pounds of rice altogether.

$93.75 - 1.75 = 92$ lb.
He had 92 pounds of rice left after losing 1.75 pounds of rice.

$92 \div 4 = 23$
He had 23 sacks of rice.

Example 4

Sandi had a length of ribbon. She used 24 centimeters of it to tie a gift and gave 1.5 meters of it to her sister. She then had 0.76 meters of ribbon left. What was the length of ribbon she had at first?

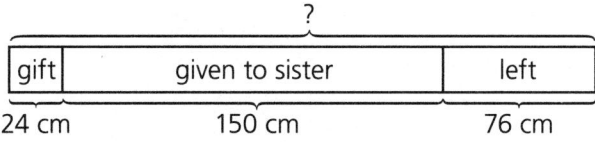

1 m = 100 cm
1.5 m = 1.5 × 100 = 150 cm
0.76 m = 0.76 × 100 = 76 cm

24 + 150 + 76 = 250
She had 250 centimeters of ribbon at first.

Example 5

A container had a weight of 6.8 kilograms when filled with Solid A. The same container had a weight of 15.4 kilograms when filled with Solid B. If Solid B was 3 times as heavy as Solid A, find the weight of the empty container.

15.4 − 6.8 = 8.6 kilograms
The weight of 2 units of Solid B is 8.6 kilograms.

2 units → 8.6 kg
1 unit → 8.6 ÷ 2
       = 4.3 kg
3 units → 3 × 4.3
       = 12.9 kg
The weight of Solid B is 12.9 kilograms.

15.4 − 12.9 = 2.5 kg
The weight of the empty container is 2.5 kilograms.

Example 6

A pencil and an eraser cost $0.70. Peter bought 3 pencils and 4 erasers and paid $2.40. Find the total cost of 2 pencils and 2 erasers.

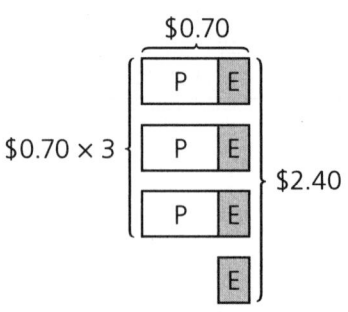

$0.70 \times 3 = $2.10
3 pencils and 3 erasers cost $2.10.

$2.40 − $2.10 = $0.30
An eraser cost $0.30.

$0.70 − $0.30 = $0.40
A pencil cost $0.40.

($0.30 × 2) + ($0.40 × 2) = $0.60 + $0.80
$\qquad\qquad\qquad\qquad\qquad\; = $1.40
2 pencils and 2 erasers cost $1.40.

Example 7

A private garage charges $1.50 for the first hour of parking and $0.60 for every additional half hour or part of a half hour. If Sam parked his car in the garage for 3 hours and 10 minutes, how much would he have to pay?

Total amount = $1.50 + (5 × $0.60)
$\qquad\qquad\quad = $1.50 + $3.00
$\qquad\qquad\quad = $4.50
He would have to pay $4.50.

Level 1 Work out the following problems.

1. Julie bought a designer handbag for $399.90. She also bought a phone case at half the price of the handbag. She then had $25.15 left. How much money did she have at first?

2. The total cost of tiling the floor of a room 12 feet by 10 feet is $256.80. What is the cost of tiling 1 square foot of the floor?

3. Mrs. Edwards used 25 pounds of flour to bake 20 loaves of bread. Estimate the amount of flour she would need for 13 loaves of bread.

4. Mrs. Anders bought some cloth. She cut 17 pieces from it, each of 2.5 yards. After that, she had 50 inches of cloth left. How much cloth did she buy at first? Give your answer in yards.

5. Find the difference in weight between 9 bags of dried corn and 5 bags of flour if one bag of dried corn has a weight of 2.25 pounds and one bag of flour has a weight of 240 ounces. Give your answer in pounds.

6. A baker had 97.5 pounds of flour. He used it to make 100 loaves of bread, each with a weight of 15 ounces. The rest of it was used to make 3 cakes. What was the weight of each cake?

7. A kilogram of crackers costs $1.25. Ms. Walters bought 25 kilograms of them. Estimate how much change she would get if she paid for the crackers with two $20 bills.

Level 2 Work out the following problems.

1. A store owner bought 50 toy cars at $1.80 each. If each toy car had cost $0.30 less, how many more toy cars could he have bought with the same amount of money?

2. A pen and an eraser cost $1.10 altogether. Alice bought 2 pens and 3 erasers for $2.55. How much did 2 pens cost?

3. A tank has a weight of 13.2 pounds when filled with Liquid A. The same tank has a weight of 21 pounds when filled with Liquid B. If Liquid B is twice as heavy as Liquid A, find the weight of the tank when empty.

4. Maria and Sally were each given the same amount of money. After Maria had spent $1.20 and Sally had spent $32.60, Maria had 3 times as much money left as Sally. Find the amount of money each girl had at first.

5. Aminah and Rasnah went to the grocer. Aminah bought 10 eggs at 30¢ each. She then bought 2 kilograms of sugar at $0.90 per kilogram. She spent $15.20 less than Rasnah altogether. If Rasnah gave the cashier a $50 bill, how much change would she get?

6. The taxi fare in a country is $2.90 for the first mile and 40¢ per mile or part of a mile for the rest of the trip. How much will Ms. Dawson have to pay to travel a distance of $8\frac{3}{4}$ miles?

7. Patty shared the cost of a party with 5 friends. She paid 20 percent of the bill, and her friends shared the balance equally. Each friend paid $15 less than Patty.

   (a) How much did Patty pay for the party?
   (b) What was the total bill?

8. Mary bought a bolt of cloth. She cut 7 equal pieces from it, each 2.5 yards long, to make some dresses. She cut another 5 pieces, each 1.5 yards long, to make some blouses. She then had $\frac{7}{12}$ of the cloth left.

How much did she pay for all the cloth if it was sold at $4.20 per yard?

9. Mr. Green bought some gloves and hats for $131.50. A hat cost $6.50 and a pair of gloves cost $3 more than a hat. If Mr. Green bought 3 more hats than pairs of gloves, how many hats and how many pairs of gloves did he buy?

10. A parking garage charges $1.20 for the first hour and $0.80 for every following half hour or part of a half hour. If Mr. Barr parks his car at the garage from 11:20 a.m. to 1:30 p.m., how much will he have to pay for parking?

Exercise 7

Percentage

Example 1

Peter spent $\frac{1}{2}$ of his savings on a car. What percentage of his savings did he spend on the car?

Method 1

$$\frac{1}{2} = \frac{50}{100}$$ (× 50 to both numerator and denominator)

= 50%

He spent 50% of his savings on the car.

Method 2

$\frac{1}{2} \times 100\%$
= 50%
He spent 50% of his savings on the car.

Example 2

0.75 of Mary's savings are coins. What percentage of her savings are coins?

0.75 × 100%
= 75%
75% of her savings are coins.

Example 3

There are 40 students in a class. 35% are boys. How many boys are there in the class?

Method 1

$100\% \rightarrow 40$ students

$1\% \rightarrow \dfrac{40}{100}$

$35\% \rightarrow 35 \times \dfrac{40}{100}$

$\qquad\quad = 14$

There are 14 boys in the class.

40 students	
35%	65%
? boys	? girls

Method 2

35% of 40

$= \dfrac{35}{100} \times 40$

$= 14$

There are 14 boys in the class.

Example 4

25% of the pieces of fruit in a basket are oranges. If there are 13 oranges in the basket, how many pieces of fruit are in the basket altogether?

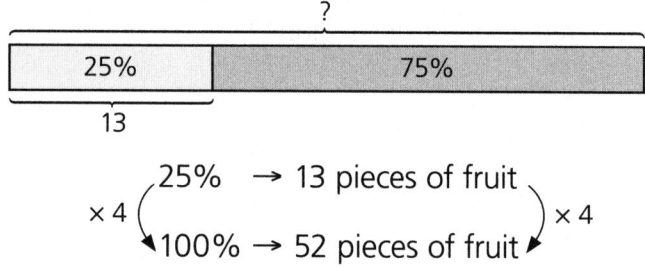

$\times 4 \begin{pmatrix} 25\% \rightarrow 13 \text{ pieces of fruit} \\ 100\% \rightarrow 52 \text{ pieces of fruit} \end{pmatrix} \times 4$

There are 52 pieces of fruit in the basket altogether.

Level 1 Work out the following problems.

1. In Jennie's class, $\frac{1}{4}$ of the students wear glasses. What percentage of the students in Jennie's class wears glasses?

2. 0.38 of the stamps in an album are foreign stamps. The rest are local stamps. If there are 124 local stamps, how many foreign stamps are there?

3. 28% of the total number of students in a school can swim. If the enrollment of the school is 1,050, how many students in the school are unable to swim?

4. 15% of the toys in a store are for babies and toddlers. If there are 225 toys for babies and toddlers, how many toys are in the store altogether?

5. Mr. Garner sold 4 bicycles to Mr. Lopez at $325 each. Mr. Lopez had to pay 7% sales tax. How much was the sales tax for the bicycles in all?

6. $\frac{3}{10}$ of the bananas in a carton are ripe. $\frac{5}{7}$ of the remaining bananas will be ripe soon. The rest of the bananas are rotten. What percentage of the bananas are rotten?

Level 2 Work out the following problems.

1. $\frac{1}{5}$ of the students in a school study French, 10% of them study Chinese, and the rest study Spanish. If there are 700 students studying Spanish in the school, how many students study Chinese?

2. 60% of a class of 40 students are to take a test. If 25% of the other students in the class go to the library, how many students would be left in the class?

3. Roy spent 40% of his money on transportation and $\frac{3}{5}$ of his remaining money on a watch. He had $108 left. How much money did he have at first?

4. The usual price of a computer and printer is $1,800. During a sale, Kathy bought a computer and printer at a discount of 15%. If she had to pay an additional 7% sales tax on the selling price, how much did she pay for the computer and printer altogether?

5. Diana bought a dress that cost $99. In addition, she had to pay 7% sales tax. How much did Diana pay in all?

6. Tara bought 250 watermelons for $1,000. She threw away 10% of them because they were rotten. She sold $\frac{2}{5}$ of the remaining watermelons at $5 each and the rest at 3 watermelons for $20. How much did she collect from the sale of the watermelons?

Exercise 8

Average

Example 1

The average number of stamps three children had was 52. If one had 72 stamps and the second child had 43 stamps, how many stamps did the third child have?

72 + 43 = 115
Two of them had 115 stamps altogether.

52 × 3 = 156
The three children had 156 stamps altogether.

156 − 115 = 41
The third child had 41 stamps.

Example 2

The average length of 3 poles is 4.2 meters. Pole A is twice as long as Pole B but 0.6 meters shorter than Pole C. Find the length of Pole C.

4.2 × 3 = 12.6 m
The total length of the 3 poles is 12.6 meters.

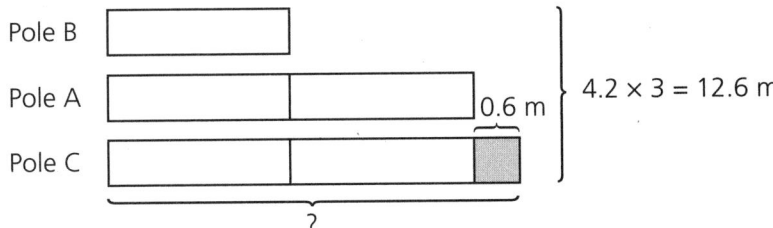

5 units → 12.6 − 0.6
        = 12 m
1 unit → 12 ÷ 5
        = 2.4 m
2 units → 2 × 2.4
        = 4.8 m

4.8 + 0.6 = 5.4 m
Pole C is 5.4 meters long.

Level 1 Work out the following problems.

1. For two years, Mr. Lee sold an average number of 55 books each month. Find the total number of books he sold in 2 years.

2. Steven earns $1,200 every two weeks. His wife earns $250 more than he does every two weeks. What is their average monthly income?

3. Loretta brought 12 oranges to school to share with Susie, Jane, and Ellen. What was the average number of oranges each girl received?

4. Three books cost $3.60 each and another two books cost $3.20 each. Find the average cost of all the books.

Level 2 Work out the following problems.

1. The sum of 6 numbers is 110. The average of 4 of the numbers is 12. Find the average of the remaining numbers.

2. A television set costs twice as much as a DVD player. If the television set costs $400, find the average cost of the two items.

3. Peter is 11 years and 8 months old. Shawn is 6 months older than Peter. Tony is 6 months younger than the average age of the 2 boys. Kevin, who is 11 years and 1 month old, joins the group.

   (a) How old is Tony?
   (b) What is the average age of the 4 boys?

4. Some boys collected an average of $80 each in donations for charity, and 28 girls collected an average of $70 each. There were 8 more girls than boys.

   (a) What was the total sum of money collected by the 2 groups?
   (b) What was the difference between the sums collected by the two groups?

Math Challenge the Singapore Way

Exercise 9

Volume

> **Notes:**
>
> ### CUBE
>
> Volume of cube (V) = Length (L) × Width (W) × Height (H)
> (Units in cm³, m³)
>
>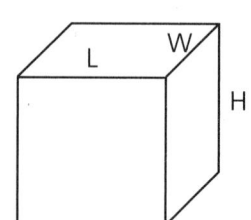
>
> A cube has the same length, width, and height, because all its edges are equal.
>
> ### CUBOID
>
> Volume of cuboid (V) = Length (L) × Width (W) × Height (H)
> (Units in cm³, m³)
>
>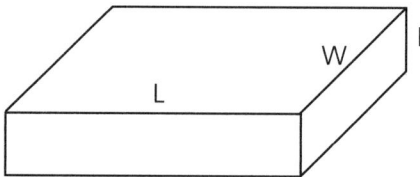
>
> ### LIQUID
>
> 1 L. = 1,000 ml
>   = 1,000 cm³

Example 1

The solid below is made up of 1-meter cubes. What is the volume of the solid?

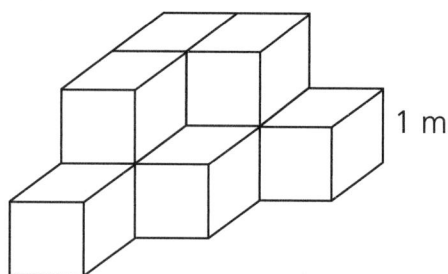

The solid is made up of 9 unit cubes.

1 × 1 × 1 = m³
9 × 1 m³ = 9 m³
The volume of the solid is 9 cubic meters.

Example 2

A rectangular tank measuring 20 centimeters by 30 centimeters by 45 centimeters is filled with water to its brim. Find the capacity of the tank in liters.

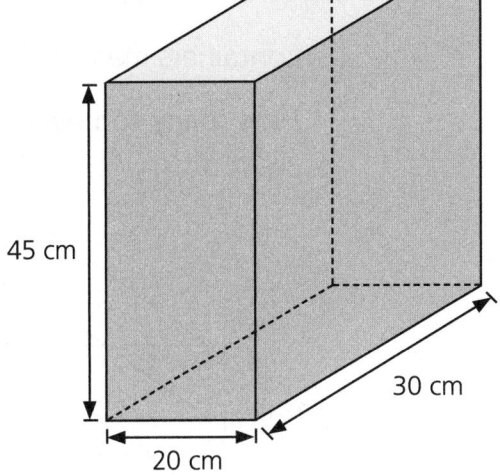

$30 \times 20 \times 45 = 27,000$ cm³
$= 27$ L.
The volume of the water in the tank is 27 liters.

Example 3

The length of a rectangular box is twice its width and three times its height. If the length is 24 centimeters, find the volume of the box.

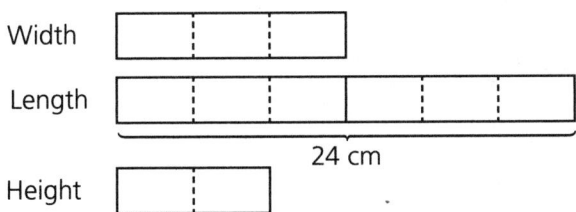

6 units → 24 cm
1 unit → 24 ÷ 6
= 4 cm
3 units → 3 × 4
= 12 cm
The width of the box is 12 centimeters.
2 units → 2 × 4
= 8 cm
The height of the box is 8 centimeters.

Volume of box = L × W × H
= 24 × 12 × 8
= 2,304 cm³
The volume of the box is 2,304 cubic centimeters.

Example 4

 A rectangular tank measuring 40 centimeters long, 20 centimeters wide, and 16 centimeters high is $\frac{1}{2}$-filled with water. All the water is poured into similar containers measuring 8 centimeters long, 4 centimeters wide, and 4 centimeters high. How many containers will be filled with water?

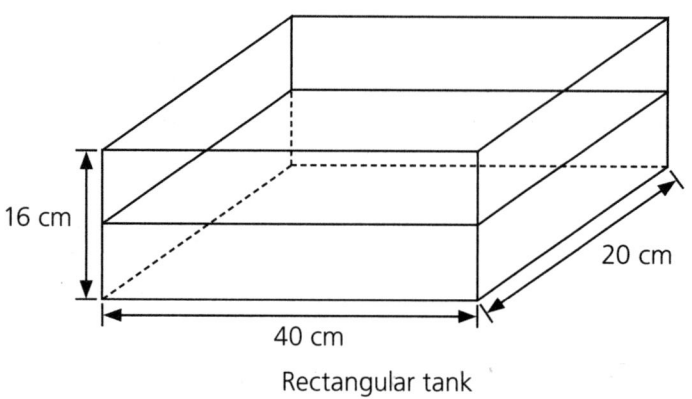
Rectangular tank

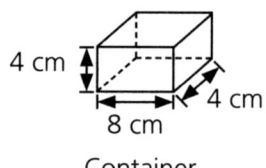
Container

Volume of water in the tank = $40 \times 20 \times 16 \times \frac{1}{2}$
= 6,400 cm³

Volume of container = $8 \times 4 \times 4$
= 128 cm³

Number of containers = 6,400 ÷ 128
= 50

50 containers will be filled with water.

Level 1 Work out the following problems.

1. The solid below is made up of 1-centimeter cubes. Find the volume of the solid.

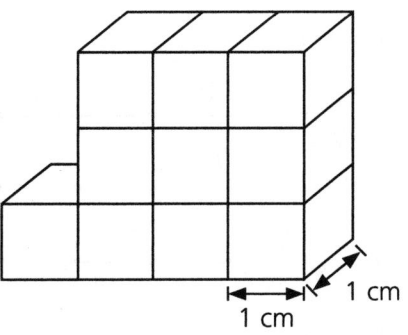

2. The solid below is made up of unit cubes. Two similar rectangular blocks are removed from the sides of the solid. How many unit cubes are left?

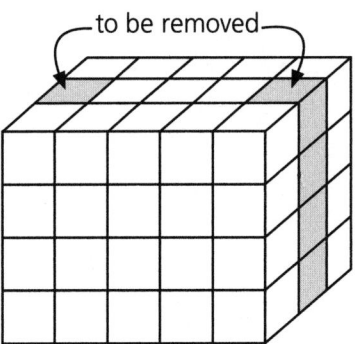

3. Draw a cuboid twice as long, twice as wide, and twice as high as the cuboid below. Then, find the volume of the cuboid you have drawn.

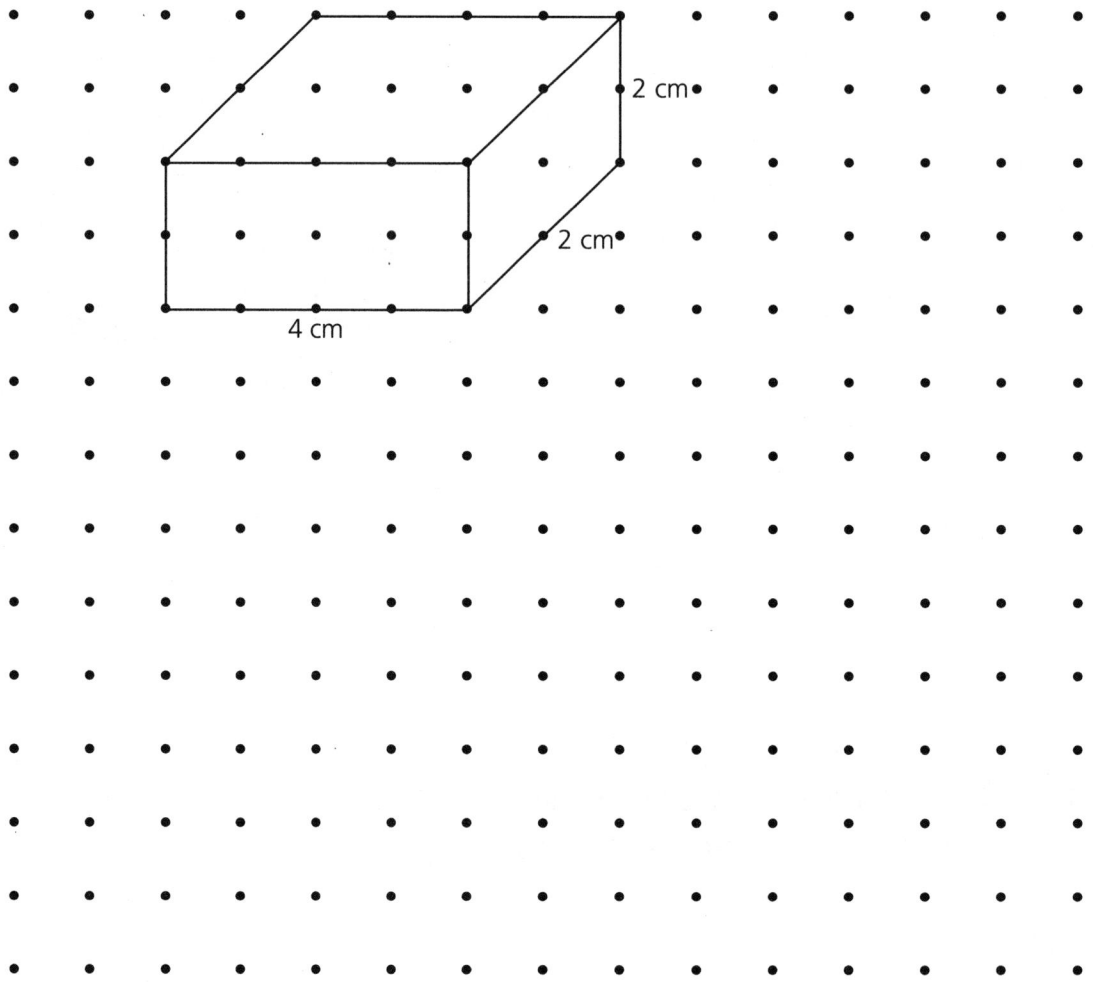

4. A rectangular tank is 15 centimeters long, 16 centimeters high, and 20 centimeters wide. How many liters and milliliters of water are in the tank?

5. Find the difference in capacity between the two containers below.

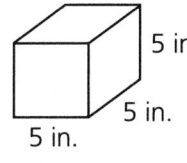

5 in.
5 in.
5 in.
Container A

5 in.
3 in.
18 in.
Container B

Level 2 Work out the following problems.

1. The 6-centimeter cube below is cut into 27 smaller cubes. What is the length of each side of the smaller cubes?

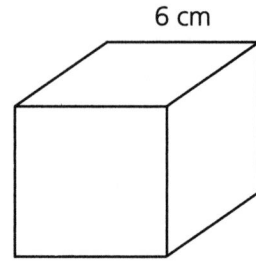

2. A rectangular box can be filled with 48 cubes. If the length of each cube is 3 inches, what is the volume of the box?

3. Kelly poured some water into an empty rectangular tank measuring 20 centimeters long and 15 centimeters wide. The height of the water level is 4 centimeters. How much water did she pour into the tank? Give your answer in liters.

4. A rectangular tank measures 32 centimeters by 20 centimeters by 12 centimeters. If it is $\frac{2}{5}$-filled with water, how much more water is needed to fill the tank completely?

5. Container X is filled completely with water. The water is then poured into Container Y. The length of Container Y is $\frac{5}{3}$ of the length of Container X. The width of Container Y is twice that of Container X. The height of Container Y is $\frac{3}{4}$ of the height of Container X. How much more water is needed to fill Container Y completely? Give your answer in liters.

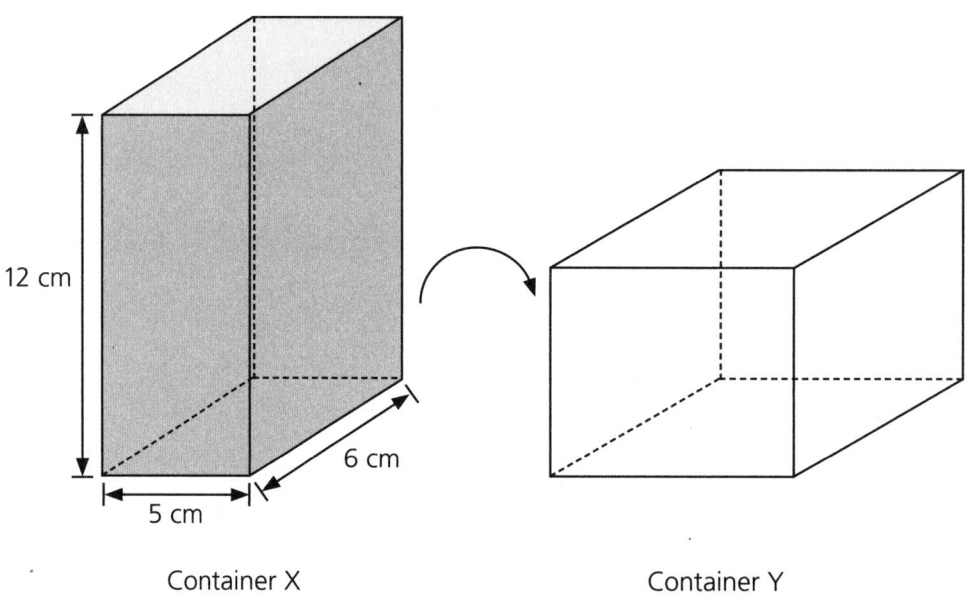

Container X

Container Y

Test Yourself 1

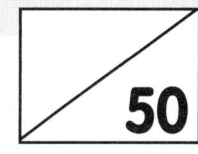

Work out the following problems.

The number of points for each question is shown in brackets [ ] at the end.

1. A store owner has 24 cartons of erasers. In each carton, there are 36 boxes, each containing 20 erasers. He intends to sell them at $3.60 per dozen. How much will he get from the sale? [5]

2. The figure below is formed using a 12-centimeter-square and a triangle. Find the area of the figure. [5]

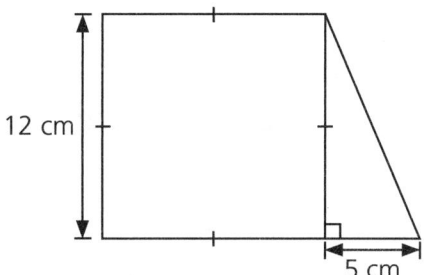

3. A rectangular can measures 50 centimeters by 30 centimeters by 20 centimeters. It is filled with water to $\frac{3}{4}$ of its height. Half the water in the can is then poured into 5 similar cans, filling them completely. What is the volume of each can? [5]

4. Mrs. Roker went to the supermarket with some money. She spent $\frac{1}{5}$ of her money on fresh fruit and vegetables, and $\frac{5}{8}$ of the remainder on 3 bottles of juice and a box of crackers. She spent the rest of the money on cereal. The box of crackers cost $\frac{1}{3}$ as much as the cereal. If she paid $12.60 for a bottle of juice, how much money did she have at first? [5]

5. Jane and Sam have $280 altogether. Sam and Rina have $340 altogether. Jane has $\frac{4}{9}$ as much money as Rina. How much money does each person have?

[5]

6. Jenny and Martin share a sum of money. If Jenny gives $\frac{1}{4}$ of her money to Martin, he will have twice the amount of money she has. Find the ratio of Jenny's original share to the total amount of money she and Martin had at first.

[5]

7. Michael gave $\frac{2}{5}$ of his salary to his parents. He spent and saved the rest in the ratio 5 : 4. If he spent $60 more than he saved, what was his salary? [5]

8. Lena and Brianna had $57.90 altogether. After Lena had spent half of her money and Brianna had spent $20.40, they each had the same amount of money left. How much did each girl have at first? [5]

9. Tim bought 2 oranges and 3 pears for $5.45. Billy bought 3 oranges and 4 pears for $7.85. Find the price of each fruit. [5]

10. Kathy spent $400 on food, $\frac{3}{10}$ of the remainder on clothes, and the rest on rent. She spent 58% of her money on food and clothes.

    (a) What percentage of the money did she spend on clothes? [3]
    (b) How much did Kathy spend altogether? [2]

Test Yourself 2

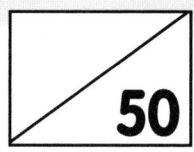

Work out the following problems.

The number of points for each question is shown in brackets [ ] at the end.

1. An empty box has a weight of 50 ounces. A man buys a certain number of small figurines, each costing $50, and places them in the box. The box now weighs 275 ounces. How much money did the man spend on the figurines if each figurine weighs 9 ounces? [5]

2. The figure below is formed using 4 identical right-angled triangles. Find the area of the figure. [5]

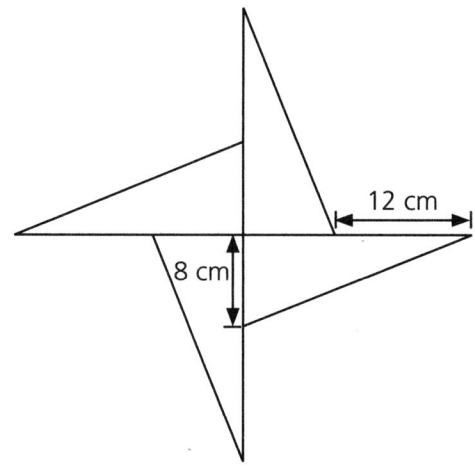

3. Complete the partly drawn cube and cuboid below. [5]

4. A container with a square base of side length 14 inches has a height of 20 inches. It is $\frac{4}{5}$-filled with water. When some of the water is poured out, the height of the water level drops by 4.5 inches. Find the amount of water that is left in the container in cubic inches. [5]

5. Marianne took 2 days to read $\frac{2}{7}$ of a book. She then took another 3 days to read the remaining 175 pages. What was the average number of pages she read per day? [5]

6. The ratio of the number of stamps Andy had to the number of stamps Brian had was 5 : 8. When Andy gave $\frac{1}{5}$ of his share to Brian, Brian had 100 stamps more than Andy. Find the number of stamps each boy had at first. [5]

7. Jason went shopping with $108.30, which was just enough to buy 2 sweaters and 1 jacket. However, he bought 1 sweater and 2 jackets, and had $11.40 left. Find the cost of the jacket. [5]

8. Michelle bought some books. The cost of $\frac{3}{4}$ of the books was $68.70. The remaining 6 books cost $4.95 each. On average, how much did she pay for each book?

[5]

9. Sue changed a $100 bill into $5 and $20 bills. The total value of the $20 bills was 4 times the value of the $5 bills. How many of each kind of bill did she have? [5]

10. Machine A makes 30 toys in 45 minutes and Machine B makes 20 toys in 30 minutes. Both machines run for 8 hours a day. If both machines are working at the same time, how many days will it take for both machines to make 4,480 toys altogether? [5]

Answers

Exercise 1 (pp.1–14)

Level 1

1. Number of pencils = 12 × 12
    = 144
   144 ÷ 10 = 14 Remainder 4
   There were 14 packets. 4 pencils were left over.

2. Number of girls = 642 − 53
    = 589
   642 + 589 = 1,231
   The enrollment of the school was 1,231.

3. Cost of toy dog = $39 + $6
    = $45
   ($39 × 2) + ($45 × 3) = $213
   The total cost of 2 toy bears and 3 toy dogs is $213.

4.
   2 units → $498
   1 unit → $498 ÷ 2 = $249
   3 units → 3 × $249 = $747
   $1000 − $747 = $253
   The change he got was $253.

5.
   2 units → $168 − $110 = $58
   1 unit → $58 ÷ 2 = $29
   The cost of the chair is $29.

6.
   6 units → $150
   1 unit → $150 ÷ 6 = $25
   The cost of each skirt is $25.

7. 4 min. → 280 words
   1 min. → 280 ÷ 4 = 70 words
   He can type 70 words in one minute.

8. 1 min. → 25 tamales
   12 min. → 12 × 25 = 300 tamales
   She can make 300 tamales in 12 minutes.

9. $2 → 5 apples
   $10 → $\frac{10}{2}$ × 5 = 25 apples
   She can buy 25 apples with $10.

Level 2

1. Amount of money store owner has
    = 350 × 40¢ = 14,000¢
   Number of rulers store owner can buy
    = 14,000¢ ÷ (40¢ − 5¢) = 400
   400 − 350 = 50
   The store owner could buy 50 more rulers with the same amount of money.

2. Cost of 3 oranges and 4 pears
    = 440¢ + 30¢ = 470¢
   Cost of 12 oranges and 16 pears
    = 4 × 470¢ = 1,880¢
   Cost of 9 pears and 12 oranges
    = 3 × 440¢ = 1,320¢
   16 − 9 = 7 pears
   7 pears → 1,880¢ − 1,320¢ = 560¢
   1 pear → 560¢ ÷ 7 = 80¢
   2 pears → 2 × 80¢ = 160¢
   2 pears cost 160¢.

3. Make a list:

Amount of water poured in liters	Amount of water in Container A in liters	Amount of water in Container B in liters
0	15	5
1	16	6
2	17	7
3	18	8
4	19	9
5	20	10 ✓

   The least amount of water poured into each container was 5 liters.

4.
   1 unit → $12 − $4 = $8
   $8 + $12 = $20
   Each boy received $20 at first.

5. Cost of 35 oz. of flour and 25 oz. of sugar = $6
    Cost of 210 oz. of flour and 150 oz.
    of sugar = 6 × $6 = $36
    Cost of 210 oz. of flour and 275 oz. of sugar = $56
    275 − 150 = 125 oz. of sugar
    125 oz. of sugar $56 − $36 = $20
    1 oz. of sugar → $\frac{20}{125}$
    25 oz. of sugar → $25 \times \frac{25}{125}$ = $4
    25 ounces of sugar cost $4.

6.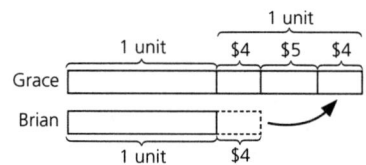

    1 unit → $4 + $5 + $4 = $13
    $13 + $4 = $17
    $13 + $4 + $5 = $22
    Grace had $22 at first. Brian had $17
    at first.

7. 30 oz. ÷ 2 = 15 oz.
    Tap B can fill a container of capacity 15 ounces
    in 1 minute.
    30 oz. + 15 oz. = 45 oz.
    Taps A and B can fill a container of capacity 45
    ounces in 1 minute.
    45 oz. → 1 min.
    30 oz. $\frac{30}{45}$ × 1 min. = $\frac{2}{3}$ × 60 sec.
                                 = 40 sec.
    It will take 40 seconds for both taps to fill the
    container at the same time.

8. (a) Cost of mailing a 60-g letter
        = 75¢ + (20¢ × 3) + 40¢ = 175¢ = $1.75
        Cost of mailing a 100-g letter
        = 75¢ + (20¢ × 3) + (40¢ × 5)
        = 335¢ = $3.35
        $1.75 + $3.35 = $5.10
        It will cost her $5.10 to mail both letters as
        2 separate pieces of mail.

    (b) 60 + 100 = 160 g
        Cost of mailing a 160-g letter
        = 75¢ + (20¢ × 3) + (40¢ × 5) + (50¢ × 6)
        = 635¢ = $6.35
        $6.35 − $5.10 = $1.25
        It will cost her $1.25 more to mail both
        letters together as 1 piece of mail.

9. Difference between the number of crackers
    eaten per day
    = 12 − 3 = 9
    Number of days Ali takes to finish his crackers
    = 45 ÷ 9 = 5 days
    Number of crackers Jorge had at first
    = (5 × 3) + 45 = 60
    Jorge had 60 crackers at first.

10. 17 units → $28,322
    1 unit → $28,322 ÷ 17
               = $1,666 (computer)
    3 units → 3 × $1,666
               = $4,998 (TV)
    $1,666 + $4,998 = $6,664
    The total cost of a TV and a computer is
    $6,664.

Exercise 2 (pp.15–25)

Level 1

1. $14\frac{2}{3} + 10\frac{1}{4} = 24\frac{11}{12}$ lb.
    They had $24\frac{11}{12}$ pounds of flour altogether.
    $24\frac{11}{12} - 20\frac{5}{6} = 4\frac{1}{12}$ lb.
    They had $4\frac{1}{12}$ pounds of flour left.

2. $\frac{3}{4} \times 8 = 6$ yd.
    He used 6 yards of cloth to make 8 pairs of
    shorts.
    $6 + 1\frac{1}{2} = 7\frac{1}{2}$ yd.
    He bought $7\frac{1}{2}$ yards of cloth.

3. $\frac{3}{5} \times 14 = 8\frac{2}{5}$ yd.
    She used $8\frac{2}{5}$ yards of cloth to make the dress.
    $14 - 8\frac{2}{5} = 5\frac{3}{5}$ yd.
    She had $5\frac{3}{5}$ yards of cloth left.

4. 3 units → $15
    1 unit → $15 ÷ 3 = $5
    5 units → 5 × $5 = $25
    She had $25 at first.

5. $6\frac{1}{3} + \frac{5}{6} = 7\frac{1}{6}$ lb.
    His brother had $7\frac{1}{6}$ pounds of sugar.
    $7\frac{1}{6} + 6\frac{1}{3} = 13\frac{1}{2}$ lb.
    They had $13\frac{1}{2}$ pounds of sugar.

6.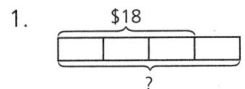

   $4 - 3 = 1$
   The difference between the number of boys and girls is 1 unit.
   1 unit → 250
   7 units → 7 × 250 = 1,750
   The enrollment of the school is 1,750.

Level 2

1. $18

   3 units → $18
   1 unit → $18 ÷ 3 = $6
   4 units → 4 × $6 = $24
   4 × $24 = $96
   He gets $96 after 4 weeks.

2.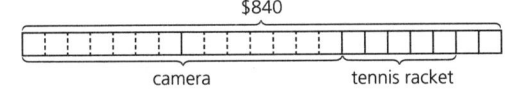

   21 units → $840
   1 unit → $840 ÷ 21 = $40
   2 units → 2 × $40 = $80
   He had $80 left.

3.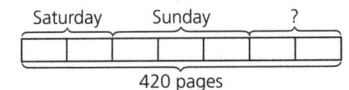

   7 units → 420 pages
   1 unit → 420 ÷ 7 = 60 pages
   2 units → 2 × 60 = 120 pages
   She had 120 pages left to read after that.

4. $400 + 350 = 750$
   $\frac{1}{5} \times 750 = 150$
   150 pieces of fruit were rotten.
   $750 - 150 - 140 - 180 = 280$
   He had 280 pieces of fruit left.

5. $\frac{5}{12} \times 120 = 50$
   Alice gave 50 postcards to Helen.
   $97 - 50 - 23 = 24$
   Jean gave 24 postcards to Helen.
   2 units → 24 postcards
   1 unit → 24 ÷ 2 = 12 postcards
   9 units → 9 × 12 = 108 postcards
   Jean had 108 postcards at first.

6. (a) $1 - \frac{1}{6} - \frac{5}{12} - \frac{7}{24} = \frac{1}{8}$
   Jack painted $\frac{1}{8}$ of the boxes green.
   $\frac{5}{12} - \frac{1}{8} = \frac{7}{24}$
   There were $\frac{7}{24}$ more red boxes than green boxes.

   (b)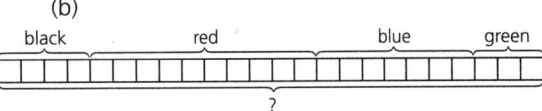

   7 units → 112 boxes
   1 unit → 112 ÷ 7 = 16 boxes
   24 units → 24 × 16 = 384 boxes
   He painted 384 boxes altogether.

7.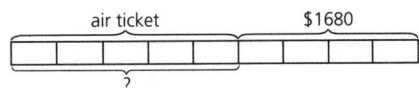

   4 units → $1,680
   1 unit → $1,680 ÷ 4 = $420
   5 units → 5 × $420 = $2,100
   He spent $2,100 to buy the airline ticket.
   $\frac{4}{5}$ of airline ticket → $2,100
   $\frac{1}{5}$ of airline ticket → $2,100 ÷ 4 = $525
   $\frac{5}{5}$ of airline ticket → 5 × $525 = $2,625
   The original price of the airline ticket was $2,625.

8. (a) $\frac{1}{3} \div 4 = \frac{1}{12}$
   She sews $\frac{1}{12}$ of the total number of garments every day.
   She would need 12 days to finish sewing all the garments.
   $12 - (4 + 3) = 5$ days
   5 days → 945 garments
   1 day → 945 ÷ 5 = 189 garments
   12 days → 12 × 189 = 2,268 garments
   She had to sew 2,268 garments altogether.

   (b) $\frac{10}{12} = \frac{5}{6}$
   She would have sewn $\frac{5}{6}$ of the total number of garments after 10 days.

Exercise 3 (pp.25–32)

Level 1

1. Area of cardboard = $\frac{40 \times 25}{2}$
   $= 500 \text{ cm}^2$
   The area of cardboard used to make the signboard is 500 square centimeters.

2. Area of triangle = $\frac{18 \times 5}{2} = 45 \text{ cm}^2$
   The area of the triangle is 45 square centimeters.

3. Area of triangle = $\frac{4 \times 6}{2} = 12 \text{ cm}^2$
   $3 \times 12 = 36 \text{ cm}^2$
   The area of triangle A is 36 square centimeters.

4. Area of paper = $\frac{8 \times 24}{2} = 96 \text{ cm}^2$
   Area of remaining paper
   $= 96 - 19 - 19 = 58 \text{ cm}^2$
   The area of the remaining piece of paper is 58 square centimeters.

Level 2

1. Length of AC = $18 \div 3 = 6$ cm
   Area of $\triangle$ ABC = $\frac{6 \times 7}{2} = 21 \text{ cm}^2$
   Area of $\triangle$ BCD = $\frac{18 \times 7}{2} = 63 \text{ cm}^2$
   Difference = $63 - 21 = 42 \text{ cm}^2$
   The difference in area between triangle CBD and triangle ABC is 42 square centimeters.

2. Since AE = EC, EC = $8 \div 2 = 4$ cm
   Area of $\triangle$ CDE = $\frac{4 \times 3}{2} = 6 \text{ cm}^2$
   The area that is not covered by the folded area. ADE is 6 square centimeters.

3. Height of triangles A and B
   $= 18 \div 2 = 9$ cm
   Area of triangles A and B
   $= 15 \times 9 = 135 \text{ cm}^2$
   Height of triangles C and D
   $= 7 \div 2 = 3.5$ cm
   Area of triangles C and D
   $= 8 \times 3.5 = 28 \text{ cm}^2$
   Total area = $135 + 28 = 163 \text{ cm}^2$
   The total area of the 4 triangles is 163 square centimeters.

*4. Since CD = DE, CD = $8 \div 2 = 4$ cm
   Area of $\triangle$ ACD = $\frac{6 \times 4}{2} = 12 \text{ cm}^2$
   Area of $\triangle$ ABE = $\frac{7}{2} \times 12 = 42 \text{ cm}^2$
   The area of triangle ABE is 42 square centimeters.

Exercise 4 (pp.33–42)

Level 1

1. $24 + 28 = 52$
   They have 52 marbles altogether.
   $28 : 52 = 7 : 13$
   The ratio of the number of marbles Jimmy has to the total number of marbles they have altogether is 7 : 13.

2. 9 units → $360
   1 unit → $360 \div 9 = \$40$
   The cost of the calculator is $40.

3. 1 unit → 6 lb.
   5 units → $5 \times 6 = 30$ lb.
   The total amount of salt in the two sacks is 30 pounds.

4. 2 units → 14
   1 unit → $14 \div 2 = 7$
   3 units → $3 \times 7 = 21$
   There are 21 boys.

5. Before:

Ivan

Ivan's brother

After:

Ivan — 10 — given to Ivan

Ivan's brother — 10

5 units → 10 comic books
1 unit → 10 ÷ 5 = 2 comic books
2 units → 2 × 2 = 4 comic books
Ivan had 4 comic books at first.

6. ($100 − $20) ÷ 2 = $40
Jane's sister has $40.
$40 + $20 = $60
Jane has $60.
40 : 60 = 2 : 3
The ratio of the amount of money her sister has to the amount of money Jane has is 2 : 3.

Level 2

1. Sara's age

Paige's age

The ratio of Paige's age to Sara's age is 3 : 2.

2. After:

Liliana

Isabel

Before:

Liliana

Isabel

2 : 2 = 1 : 1
The ratio of Isabel's share to Liliana's share at first was 1 : 1.

3. Before:

Mr. Lilliard

Mr. Sanders

After: ? $2,310

Mr. Lillard

Mr. Sanders
?

11 units → $2,310
1 unit → $2,310 ÷ 11 = $210
5 units → 5 × $210 = $1,050
8 units → 8 × $210 = $1,680
Mr. Lillard's salary in April is $1,050.
Mr. Sanders's salary in April is $1,680.

4. Before:

Boys

Girls

After:

Boys

Girls
22

1 unit → 22 members
7 units → 7 × 22 = 154 members
There were 154 members in the club last year.

5. $135,696

Wife

Four sons
?

3 units → $135,696
1 unit → $135,696 ÷ 3 = $45,232
5 units → 5 × $45,232 = $226,160
$226 160 ÷ 4 = $56,540
Each son received $56,540.

6. ? 24

Number of oranges left unsold in Stall A

Number of oranges left unsold in Stall B
?

4 units → 24 oranges
1 unit → 24 ÷ 4 = 6 oranges
3 units → 3 × 6 = 18 oranges (Stall A)
7 units → 7 × 6 = 42 oranges (Stall B)
42 − 6 = 36 oranges
18 : 36 = 1 : 2
The new ratio of the number of oranges left in Stall A to those in Stall B is 1 : 2.

7.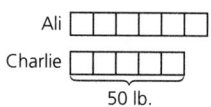

5 units → 50 lb.
1 unit → 50 ÷ 5 = 10 lb.
6 units → 6 × 10 = 60 lb.
Ali's weight is 60 pounds.

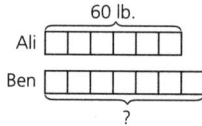

6 units → 60 lb.
1 unit → 60 ÷ 6 = 10 lb.
7 units → 7 × 10 = 70 lb.
Ben's weight is 70 pounds.

 8.

3 units → 172 + 462 + 65
       = 699 marbles
1 unit → 699 ÷ 3 = 233 marbles
2 units → 2 × 233 = 466 marbles
He had 466 red marbles at first.

Exercise 5 (pp.43–49)

Level 1

1. $\angle a = 180° - 62° - 77° = 41°$

2. $\angle b = 360° - 100° - 90° - 15° = 155°$

3. $\angle c = 180° - 55° - 20° = 105°$

4. (a) All sides of a rhombus are equal.
       Perimeter = 4 × 10 = 40 cm
       The perimeter of the figure is 40 centimeters.

   (b) $\angle DCB = 45°$ (Opposite angles in a rhombus)

5. (a) PQRS is a parallelogram.
       Therefore, $\angle SPQ + \angle PSR = 180°$
       The value of the sum of $\angle SPQ$ and $\angle PSR$ is 180°.

   (b) $\angle SPQ = 65°$ (Opposite angles in a parallelogram)

6. $\angle ZWX = 180° - 115° = 65°$ (Angles between parallel sides)

7.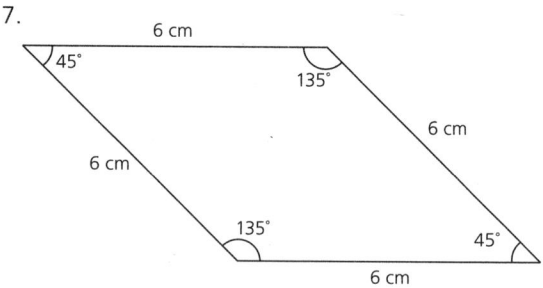

8.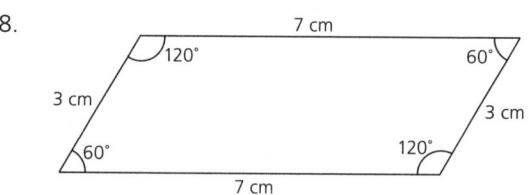

9.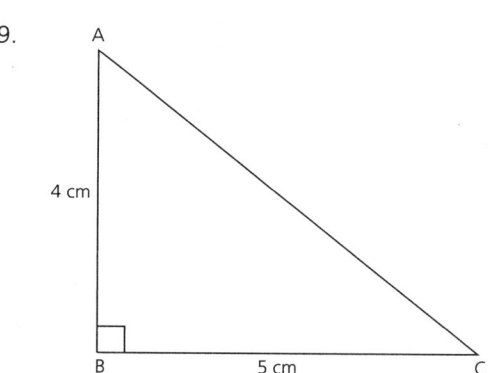

10. $\angle y = 180° - (50° \times 2) = 80°$

11. $\angle CAB = 132° - 60° = 72°$
    $\angle ABC = 180° - (72° \times 2) = 36°$

*12. $\angle PQR = \angle RPQ = (180° - 40°) \div 2$
                              $= 70°$
    $\angle QST = 180° - 70° = 110°$
    $\angle c = 110° - 20° = 90°$
    $\angle SPQ = 180° - 20° - 70° = 90°$
    $\angle d = 90° - 70° = 20°$

*13. $\angle TOU = 180° - 105° = 75°$
    $\angle q = 180° - 75° - 75° = 30°$
    $\angle s = 180° - 60° = 120°$

Exercise 6 (pp.50–64)

Level 1

1. Price of phone case = $399.90 ÷ 2
   = $199.95
   $399.90 + $199.95 + $25.15
   = $625
   She had $625 at first.

2. Area of room = 12 × 10 = 120 square feet
   $256.80 ÷ 120 = $2.14
   The cost of tiling 1 square foot of the floor is $2.14.

3. 25 ÷ 20 = 1.25 lb.
   1.25 pounds of flour is used to bake a loaf of bread.
   13 × 1.25 ≈ 13 × 1 = 13 lb.
   She would need about 13 pounds of flour to bake 13 loaves of bread.

4. 17 × 2.5 = 42.5 yd.
   1 yd. = 3 ft. = 3 × 12 in. = 36 in.
   27 ÷ 36 = 0.75 yd.
   42.5 + 0.75 = 43.25 yd.
   She bought 43.25 yards of cloth at first.

5. Weight of 9 bags of dried corn = 9 × 2.25 = 20.25 lb.
   1 lb. = 16 oz.
   Weight of 5 bags of flour = 5 × 240 = 1,200 oz.
   1,200 ÷ 16 = 75 lb.
   Difference in weight = 75 − 20.25 = 54.75 lb.
   The difference in weight between 9 bags of dried corn and 5 bags of flour is 54.75 pounds.

6. Weight of 100 loaves of bread = 100 × 15 = 1,500 oz.
   1 lb. = 16 oz.
   1,500 ÷ 16 = 93.75 lb.
   97.5 − 93.75 = 3.75 lb.
   3.75 ÷ 3 = 1.25 lb.
   The weight of each cake was 1.25 pounds.

7. Cost of 25 kg of crackers
   = 25 × $1.25 = $31.25
   2 × $20 = $40
   $40 − $31.25 ≈ $40 − $31
   = $9
   She would get about $9 change.

Level 2

1. $0.30 × 50 = $15
   $1.80 − $0.30 = $1.50
   $15 ÷ $1.50 = 10
   He could have bought 10 more toy cars with the same amount of money.

2. Cost of 2 pens and 2 erasers
   = $1.10 × 2 = $2.20
   Cost of an eraser = $2.55 − $2.20
   = $0.35
   Cost of 2 pens = $2.55 − (3 × $0.35)
   = $1.50
   2 pens cost $1.50.

3. Difference in weight between Liquids A and B
   = 21 − 13.2 = 7.8 lb.

   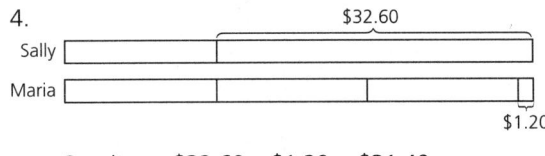
   7.8 lb.

   Weight of empty tank = 13.2 − 7.8
   = 5.4 lb.
   The weight of the empty tank is 5.4 pounds.

4.
   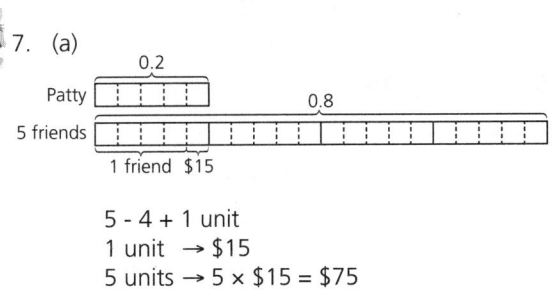

   2 units → $32.60 − $1.20 = $31.40
   1 unit → $31.40 ÷ 2 = $15.70
   $15.70 + $32.60 = $48.30
   Each girl had $48.30 at first.

5. Cost of 10 eggs = 10 × 30¢ = 300¢
   = $3
   Cost of 2 kg of sugar = 2 × $0.90
   = $1.80
   Amount Aminah spent = $3 + $1.80
   = $4.80
   Amount Rasnah spent
   = $4.80 + $15.20 = $20
   $50 − $20 = $30
   She would get $30 in change.

6. $8\frac{3}{4}$ mi. ≈ 9 mi.
   Taxi fare = (1 × $2.90) + (8 × $0.40)
   = $6.10
   Ms. Dawson will have to pay $6.10.

7. (a)

   5 − 4 + 1 unit
   1 unit → $15
   5 units → 5 × $15 = $75
   Patty paid $75 for the party.

   (b) 5 × 5 = 25 units
   25 units → 25 × $15 = $375
   The total bill was $375.

8. $(2.5 \times 7) + (1.5 \times 5) = 25$ yd.
   She made dresses and blouses with 25 yards of cloth.
   $1 - \frac{7}{12} = \frac{5}{12}$
   She used $\frac{5}{12}$ of the cloth.
   $\frac{5}{12}$ of the cloth → 25 yd.
   $\frac{1}{12}$ of the cloth → 25 ÷ 5
   = 5 yd.
   $\frac{12}{12}$ of the cloth → 12 × 5
   = 60 yd.
   She bought 60 yards of cloth.
   60 × $4.20 = $252
   She paid $252 for all the cloth.

9. Cost of a pair of gloves
   = $6.50 + $3 = $9.50

   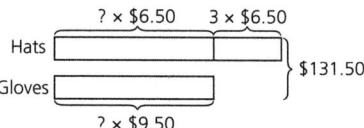

   Number of pairs of gloves bought
   = ($131.50 − 3 × $6.50) ÷ ($6.50 + $9.50)
   = 7
   Number of hats bought = 7 + 3 = 10
   Mr. Green bought 10 hats and 7 pairs of gloves.

10. Make a list:

Time	Cost
11:20 a.m. to 12:20 p.m.	$1.20
12:20 p.m. to 12:50 p.m.	$0.80
12:50 p.m. to 1:20 p.m.	$0.80
1:20 p.m. to 1:30 p.m.	$0.80

    Total parking fee
    = $1.20 + (3 × $0.80) = $3.60
    He will have to pay $3.60 for parking.

Exercise 7 (pp.65–72)

Level 1
1. Percentage = $\frac{1}{4} \times 100\% = 25\%$
   25% of the students in Jennie's class wear glasses.

2. $0.38 = \frac{38}{100} = 38\%$
   100% − 38% = 62%
   62% → 124 stamps
   1% → 124 ÷ 62 = 2 stamps
   38% → 38 × 2 = 76 stamps
   There are 76 foreign stamps.

3. 100% − 28% = 72%
   72% of the students are unable to swim.
   100% → 1,050
   1% → 1,050 ÷ 100 = 10.5
   72% → 72 × 10.5 = 756
   756 students are unable to swim.

4. 15% → 225 toys
   1% → 225 ÷ 15 = 15 toys
   100% → 100 × 15 = 1,500
   There are 1,500 toys in the store altogether.

5. Cost of 4 bicycles = 4 × $325
   = $1,300
   7% sales tax = 7% × $1,300 = $91
   The sales tax for the bicycles was $91 in all.

6.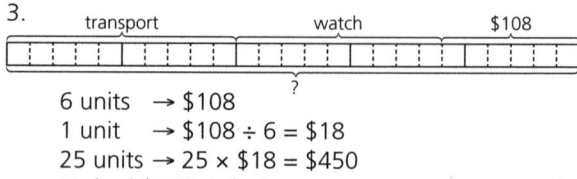

   10 units → 100%
   1 unit → 100% ÷ 10 = 10%
   2 units → 2 × 10% = 20%
   20% of the bananas are rotten.

Level 2
1. $\frac{1}{5} \times 100\% = 20\%$
   20% of the students study French.
   100% − 20% − 10% = 70%
   70% of the students study Spanish.
   70% → 700 students
   1% → 700 ÷ 70 = 10 students
   10% → 10 × 10 = 100 students
   100 students study Chinese.

2. 100% − 60% = 40%
   40% × 40 = 16 students
   16 students are not to take a test.
   25% × 16 = 4 students
   4 students went to the library.
   40 − 4 = 36 students
   There would be 36 students left in the class.

3.
   | transport | watch | $108 |

   6 units → $108
   1 unit → $108 ÷ 6 = $18
   25 units → 25 × $18 = $450
   He had $450 at first.

4. 100% − 15% = 85%
   Kathy paid 85% of the usual price.
   85% × $1,800 = $1,530
   7% × $1,530 = $107.10
   $1,530 + $107.10 = $1,637.10
   Kathy paid $1,637.10 for the computer and printer altogether.

5. 7% × $99 = $6.93
   $99 + $6.93 = $105.93
   Diana paid $105.93 in all.

6. 100% − 10% = 90%
   90% × 250 = 225
   225 watermelons were not rotten.
   $\frac{2}{5} \times 225 = 90$
   90 watermelons were sold at $5 each.
   225 − 90 = 135
   135 watermelons were sold at 3 for $20.
   ($5 × 90) + (135 ÷ 3) × $20 = $1,350
   She collected $1,350 from the sale of the watermelons.

Exercise 8 (pp.73–77)

Level 1

1. 55 × 12 × 2 = 1,320
   The total number of books he sold in 2 years is 1,320.

2. $1,200 + $250 = $1,450
   His wife earns $1,450 every two weeks.
   $1,200 + $1,450 = $2,650
   $2,650 × 2 = $5,300
   They earn $2,650 every two weeks.
   $2,650 × 2 = $5,300
   Their average monthly income is $5,300.

3. 12 ÷ 4 = 3
   The average number of oranges each girl received was 3.

4. ($3.60 × 3) + ($3.20 × 2) = $17.20
   $17.20 ÷ 5 = $3.44
   The average cost of all the books is $3.44.

Level 2

1. 12 × 4 = 48
   110 − 48 = 62
   62 ÷ 2 = 31
   The average of the remaining numbers is 31.

2. $400 ÷ 2 = $200
   The DVD player costs $200.
   $400 + $200 = $600
   The total cost is $600.
   $600 ÷ 2 = $300
   The average cost of the two items is $300.

3. (a) 11 years and 8 months + 6 months = 12 years and 2 months
   Shawn is 12 years and 2 months old.
   12 years and 2 months + 11 years and 8 months = 23 years and 10 months
   Average age of the 2 boys = (23 years and 10 months) ÷ 2 = 11 years and 11 months
   11 years and 11 months − 6 months = 11 years and 5 months
   Tony is 11 years and 5 months old.

   (b) Total age of the 4 boys = 23 years and 10 months + 11 years and 5 months + 11 years and 1 month = 46 years and 4 months
   Average age = (46 years and 4 months) ÷ 4 = 11 years and 7 months
   The average age of the 4 boys is 11 years and 7 months.

4. (a) 28 − 8 = 20
   There were 20 boys.
   20 × $80 = $1,600
   The boys collected $1,600.
   28 × $70 = $1,960
   The girls collected $1,960.
   $1,600 + $1,960 = 3,560
   The total sum of money collected by the 2 groups was $3,560.

   (b) $1,960 − $1,600 = $360
   The difference between the sums collected by the 2 groups was $360.

Exercise 9 (pp.78–86)

Level 1

1. Volume of 1 cube = 1 × 1 × 1 = 1 cm³
   Volume of the solid = 10 × 1 = 10 cm³
   The volume of the solid is 10 cubic centimeters.

2. Number of unit cubes left
   = (5 × 3 × 4) − (2 × 4) = 52
   52 unit cubes are left.

3.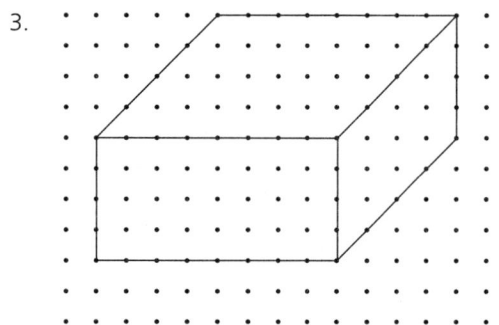

Volume of cuboid = 8 × 4 × 4 = 128 cm³
The volume of the cuboid drawn is 128 cubic centimeters.

4. 15 × 16 × 20 = 4,800 cm³
4,800 cm³ = 4 L 800 ml
There are 4 liters, 800 milliliters in the tank.

5. Volume of Container A = 5 × 5 × 5
   = 125 cubic inches
   Volume of Container B = 18 × 5 × 3
   = 270 cubic inches
   270 − 125 = 145 cubic inches
   The difference in capacity between the two containers is 145 cubic inches.

Level 2

1. Volume of the cube = 6 × 6 × 6
   = 216 cm³
   The volume of the cube is 216 cubic centimeters.
   Volume of 1 smaller cube = 216 ÷ 27 = 8 cm³
   2 × 2 × 2 = 8 cm³
   The length of each side of the smaller cubes is 2 centimeters.

2. Volume of each cube = 3 × 3 × 3
   = 27 cubic inches
   48 × 27 = 1,296 cubic inches
   The volume of the box is 1,296 cubic inches.

3. Volume of water = 20 × 15 × 4
   = 1,200 cm³ = 1.2 L
   She poured 1.2 liters of water into the tank.

4. Volume of water needed
   = 32 × 20 × 12 × $\frac{3}{5}$
   = 4,608 cm³ = 4,608 ml = 4 L 608 ml
   4 liters and 608 milliliters of water are needed to fill the tank completely.

5. Volume of water in Container X
   = 6 × 5 × 12 = 360 cm³
   Volume of Container Y
   = (6 × $\frac{5}{3}$) × (5 × 2) × (12 × $\frac{3}{4}$)
   = 10 × 10 × 9 = 900 cm³

Amount of water needed
= 900 − 360 = 540 cm³ = 540 ml = 0.54 L.
0.54 liters of water are needed to fill Container Y completely.

Test Yourself 1 (pp.87–91)

1. Number of erasers in 24 cartons
   = 24 × 36 × 20 = 17,280
   Number of dozens of erasers
   = 17,280 ÷ 12 = 1,440
   1,440 × $3.60 = $5,184
   He will get $5,184 from the sale.

2. Area of figure = Area of square + Area
                                      of triangle
   = (12 × 12) + ($\frac{12 \times 5}{2}$)
   = 144 + 30 = 174 cm²
   The area of the figure is 174 square centimeters.

3. Amount of water in can
   = 50 × 30 × 20 × $\frac{3}{4}$ = 22,500 cm³
   Amount of water poured
   = 22,500 ÷ 2 = 11,250 cm³
   Volume of each can = 11,250 ÷ 5
   = 2,250 cm³
   The volume of each can is 2,250 cm³.

4.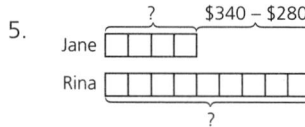

   4 units → 3 × $12.60 = $37.80
   1 unit → $37.80 ÷ 4 = $9.45
   10 units → 10 × $9.45 = $94.50
   She had $94.50 at first.

5.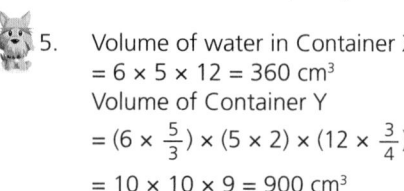

   5 units → $340 − $280 = $60
   1 unit → $60 ÷ 5 = $12
   4 units → 4 × $12 = $48 (Jane)
   9 units → 9 × $12 = $108 (Rina)
   $280 − $48 = $232 (Sam)
   Jane has $48, Rina has $108, and Sam has $232.

6.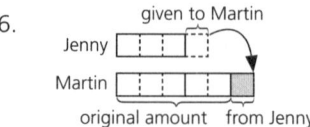

   Total number of units = 3 + 6 = 9
   The ratio of Jenny's original share to the total amount of money at first is 4 : 9.

7.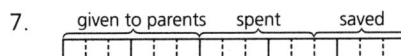
Difference between the amount spent and saved = 5 − 4 = 1 unit
1 unit → $60
15 units → 15 × $60 = $900
His salary was $900.

8.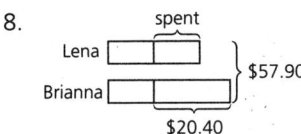
3 units → $57.90 − $20.40 = $37.50
1 unit → $37.50 ÷ 3 = $12.50
2 units → 2 × $12.50 = $25 (Lena)
$12.50 + $20.40 = $32.90 (Brianna)
Lena had $25 at first and Brianna had $32.90 at first.

9.
Cost of 2 oranges and 3 pears = $5.45
Cost of 6 oranges and 9 pears = 3 × $5.45 = $16.35
Cost of 3 oranges and 4 pears = $7.85
Cost of 6 oranges and 8 pears = 2 × $7.85 = $15.70
Cost of a pear = $16.35 − $15.70 = $0.65
Cost of 2 oranges = $5.45 − (3 × $0.65) = $5.45 − $1.95 = $3.50
Cost of an orange = $3.50 ÷ 2 = $1.75
The cost of a pear is $0.65, and the cost of an orange is $1.75.

10. (a)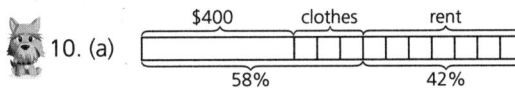
7 units → 42%
1 unit → 42% ÷ 7 = 6%
3 units → 3 × 6% = 18%
She spent 18% of the money on clothes.

(b) 58% − 18% = 40%
40% → $400
1% → $400 ÷ 40 = $10
100% → 100 × $10 = $1,000
Kathy spent $1,000 altogether.

Test Yourself 2 (pp.92–96)

1. Total weight of the figurines = 275 − 50 = 225 oz.
Number of figurines = 225 oz. ÷ 9 oz. = 25
Total money spent on the figurines = 25 × $50 = $1,250
The man spent $1,250 on the figurines.

2. Height of each triangle = 12 + 8 = 20 cm
Area of 1 triangle = $\frac{20 \times 8}{2}$ = 80 cm²
Area of the figure = 4 × 80 = 320 cm²
The area of the figure is 320 square centimeters.

3.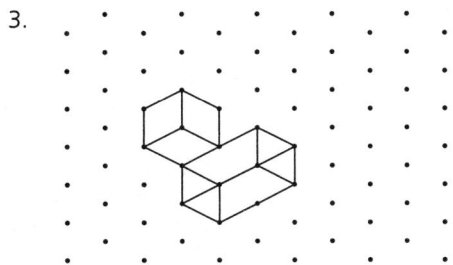

4. Original height of water level = $\frac{4}{5}$ × 20 = 16 in.
New height of water level = 16 − 4.5 = 11.5 in.
Amount of water left in container = 14 × 14 × 11.5 = 2,254 cubic inches
The amount of water left in the container is 2,254 cubic inches.

5.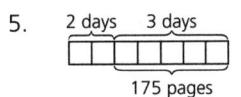
5 units → 175 pages
1 unit → 175 ÷ 5 = 35 pages
7 units → 7 × 35 = 245 pages
245 ÷ 5 = 49 pages
She read an average of 49 pages per day.

6.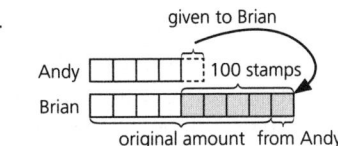
5 units → 100 stamps
1 unit → 100 ÷ 5 = 20 stamps
8 units → 8 × 20 = 160 stamps
Andy had 100 stamps at first and Brian had 160 stamps at first.

7. The cost of a sweater is $11.40 more than the cost of a jacket.

Cost of sweater		$11.40	
Cost of sweater		$11.40	} $108.30
Cost of jacket			

3 units → $108.30 − (2 × $11.40)
= $108.30 − $22.80 = $85.50
1 unit → $85.50 ÷ 3 = $28.50
The cost of the jacket is $28.50.

8. 1 unit → 6 books
4 units → 4 × 6 = 24 books
Total cost of the remaining 6 books
= 6 × $4.95 = $29.70
Total cost of the books
= $29.70 + $68.70 = $98.40
Average cost = $98.40 ÷ 24 = $4.10
On average, she paid $4.10 for each book.

9.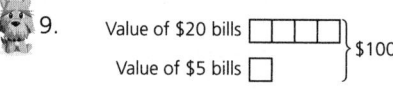

5 units → $100
1 unit → $100 ÷ 5 = $20
4 units → 4 × $20 = $80
Number of $5 bills = $20 ÷ $5 = 4
Number of $20 bills = $80 ÷ $20
= 4
She had 4 $5 bills and 4 $20 bills.

10. Number of toys Machine A makes in
1 hr. = $\frac{30}{45} \times 60 = 40$

Number of toys Machine B makes in
1 hr. = $\frac{20}{30} \times 60 = 40$

Number of toys Machine A and B
make in 1 hr. = 40 + 40 = 80
Number of hours needed to make
4,480 toys = 4,480 ÷ 80 = 56 hours
Number of days = 56 ÷ 8 = 7 days
Both machines will take 7 days to make 4,480 toys altogether.

Notes

Notes

Notes

Notes

Notes

Notes